Cambridge County  *Geographies*

Scotland

General Editor: W. Murison, M.A.

STIRLINGSHIRE

Cambridge County Geographies

STIRLINGSHIRE

by

W. DOUGLAS SIMPSON, D.Litt.

Librarian in the University of Aberdeen

With Maps, Diagrams and Illustrations

CAMBRIDGE

AT THE UNIVERSITY PRESS

1928

CAMBRIDGE UNIVERSITY PRESS
Cambridge, New York, Melbourne, Madrid, Cape Town,
Singapore, São Paulo, Delhi, Mexico City

Cambridge University Press
The Edinburgh Building, Cambridge CB2 8RU, UK

Published in the United States of America by Cambridge University Press, New York

www.cambridge.org
Information on this title: www.cambridge.org/9781107671621

© Cambridge University Press 1928

First published 1928
First paperback edition 2013

A catalogue record for this publication is available from the British Library

ISBN 978-1-107-67162-1 Paperback

PREFATORY NOTE

M Y thanks are due to numerous correspondents who have kindly favoured me with information upon points of detail; and specially to Mr David B. Morris, Town Clerk of Stirling, who has placed his great local knowledge unreservedly at my disposal, and has added to my indebtedness by revising the whole manuscript.

W. DOUGLAS SIMPSON

King's College
University of Aberdeen
1928

CONTENTS

ILLUSTRATIONS

MAPS AND PLANS

The illustrations on pp. 18, 92, 94, 104, 106, 107, 115, 129, 131, 133 are from photographs by Messrs Valentine and Son; those on pp. 30, 72, 74, 75, 76, 77, 78, 85 (lower figure), 86, 98 are reproduced by kind permission of the Society of Antiquaries of Scotland; that on p. 37 is from Dr Jas. Ritchie's *Influence of Man on Animal Life in Scotland*; that on p. 39 from A. H. Evans' *Birds of Britain*. The illustrations on pp. 55, 56, 57 were supplied by the Carron Company, Falkirk; that on p. 67 is from P. Hume Brown's *History of Scotland*; that on p. 79 from Cochran-Patrick's *Medieval Scotland*; that on p. 85 (upper figure), from Dr G. Macdonald's *The Roman Wall in Scotland*; those on pp. 90, 101, 103 are from photographs supplied by F. W. Hardie; that on p. 123 from a photograph by T. and R. Annan and Sons; and the Curator of the Smith Institute, Stirling, kindly supplied those on pp. 135 and 136.

1. County and Shire. The Origin of Stirling.

The division of Scotland, for administrative purposes, into parishes and counties, or shires, was the work of the Anglo-Norman barons and churchmen who thronged into the country during the twelfth and thirteenth centuries, and reorganised her institutions upon a feudal basis. The word "shire" is derived from the Old English "scir," office, administration. "County" is from the corresponding Norman-French word, "comté,"Latin "comitatus," strictly the territorial appanage of a feudal earl or count. In Old English times the administrative officer of the shire was known as the "shire-reeve" or sheriff; and both the office and the name were preserved by the Norman conquerors. As the latter to a large extent availed themselves of existing political divisions and institutions, the Old English shire was in many cases conferred as a county upon a Norman earl, in whom the sheriffdom was vested. Thus in course of time the terms shire and county came to be interchangeable.

Very little is known as to the origin of the Scottish shires, but it seems certain that some of them embody territorial units dating back to early Celtic times. Still less is known as to the principles which determined their boundaries. In many cases political considerations now forgotten, and manorial boundaries of the early proprietors, may have played their part. In other cases the county has obviously come in existence as an administrative area in dependence upon some centre, such as a royal castle. This is the case with Stirlingshire, which has grown up round the fortified rock that from the earliest ages sentinelled the passage of the Forth. Situated at the most inland point where a bridge could be thrown across the estuary, Stirling was always a place of vital strategic importance. Everyone travelling from Lowlands to Highlands, or *vice versa*, had to pass by this route. In modern times its importance has been much diminished by the construction of alternative railway routes by the Forth Bridge and at Alloa: but until last century the predominance of Stirling Bridge remained unchallenged. With the growth of motor traffic the road across the bridge at Stirling is again one of the busiest in the country. For this reason it has been well said that "Stirling, like a huge brooch, clasps Highlands and Lowlands together." But the position of Stirling has a significance wider than that of its relation to the passage of the Forth. It is not merely a "bridge-town," it is a "gap-town." In the long range of hills which forms the southern boundary of Strathmore, comprising the Sidlaws, the Ochils, and the Lennox Hills, two important gaps occur, both of which have been used as routes from early times. One gap marks the passage of the Tay between the Sidlaws and the Ochils, and was guarded by the walled town of Perth. The second gap is formed by the valley of the Forth passing between the Ochils and the Lennox Hills. This gap was sentinelled by the castled rock of Stirling.

Owing to the fact that they grew up under political conditions long obsolete, the marches of the Scottish counties were often strangely anomalous until rectified by a Commission in 1891. Previous to that date the boundary of Stirlingshire showed some remarkable peculiarities. The parish of Logie, north of the Forth, was intersected by a strip of Perthshire, while the parish of Alva, now in Clackmannan, was formerly an outlying part of Stirling. A portion of the Perthshire parish of Lecropt was also in Stirling. In the parish of Kippen, Perth overstepped the Forth and included two strips on the Stirling side. A portion of the Dumbarton parish of New Kilpatrick, containing Milngavie, was included in Stirling.

The origin of the name Stirling is doubtful. Old forms are Estrevelyn, Striviling, and Struelin. The term Snawdoun, or Snowdon, often given to Stirling by medieval chroniclers, is perhaps a corruption of Gaelic words meaning the fortress with a well.

2. General Characteristics.

Geologically and geographically, Scotland may be divided into three districts, bounded by two great "faults," or lines of fracture in the earth's crust, which traverse the country from north-east to south-west, one running from St Abb's Head to Girvan, the other from Stonehaven to Helensburgh. The three districts thus formed are known as the Highlands, the Central Lowlands, and the Southern Uplands; and each has its characteristic physical, biological, and industrial conditions. Stirlingshire lies mainly in the Central Lowlands, but its north-west portion, along the east shore of Loch Lomond, is included in the Highland area. Moreover the main mass of the county, though belonging geologically to the Central Lowlands, is anything but uniformly Lowland in character. The plain country is in fact confined to the valleys of the Forth with its tributaries the Allan, Bannock, Carron, and Avon, the Kelvin and a small strip in the valleys of the Endrick and the Blane near Drymen and Killearn, at the lower end of Loch Lomond. In the centre of the county rises the great volcanic area of the Lennox Hills, and intrusive masses such as the Castle Rock of Stirling and the Abbey Craig diversify the scenery of the Lowland strips. Thus Stirlingshire contains within its limits almost every variety of Scottish scenery, from the fertile haughs of the Carseland to the treeless slopes of the Lennox Hills and the wild mountain ridges of Ben Lomond.

These physical differences between the various parts of Stirlingshire have exercised important effects upon its flora, its fauna, and its human life. The Highland area is a district of bare mountains and open moor, scantily wooded and supporting a thin population. Somewhat similar conditions prevail in the central volcanic hills, the grassy slopes of which are mainly devoted to pasture. But the plain country is an area of dense population, owing partly to its fertility, but mainly to its mineral resources. Stirlingshire is thus partly an agricultural and partly an industrial county. Even its agriculture is largely industrialised, for the farmer in Stirlingshire no longer grows his produce to meet his own needs, as his forebears did, but to sell in the open market. There is also much pastoral ground, and a considerable area is waste or given over to sport. There are extensive grouse moors on the Lennox and Kilpatrick Hills.

The industrial district of Stirlingshire coincides with the great eastern coal-field. This is one of the most densely populated areas in Scotland. Its prosperity depends on a series of coincidences : (1) the presence of iron and coal in close association; (2) the neighbourhood of limestone and gannister, used in smelting; (3) the proximity, in both directions, of the sea, with good ports. The ease of communications east and west with these seaports, owing to the narrowness of Scotland between the two estuaries, Forth and Clyde, has greatly aided the industrial progress of Stirlingshire. In other directions also the physical features and natural conditions of the county have closely affected its industrial development. For example, the woollen industry, still plied at Stirling and Bannockburn, originated through the presence locally of good pasture land in the Ochil and Lennox Hills, and abundant water-power in the streams which hurry down from the heights ; while, at a later stage, the exploitation of the coal-fields replaced the older water-power by other sources of energy dependent on cheap fuel.

Although the eastern corner of Stirlingshire skirts the estuary of the Forth, the county is in no sense a maritime one, and its fisheries are of small importance. Yet it contains one port of consequence, Grangemouth, which is the eastern end of the Forth and Clyde canal, and also serves as a seaward outlet for the great iron and coal area which adjoins. From this port a large coastwise and foreign trade is carried on; quantities of coal being exported, while pig-iron is imported from Middlesborough, most of which is forwarded to Glasgow. There is also a considerable ship-building industry at this port.

3. Shape, Size, Boundaries.

Stirlingshire lies between 55° 53′ and 56° 19′ north latitude, and between 3° 37′ and 4° 42′ west longitude. The longest straight line that can be drawn in the shire extends from the boundary west of Linlithgow to the shore of Loch Lomond opposite Tarbert Pier. This line measures 43 miles. The greatest breadth of the shire occurs on a line running north and south between the point where Clackmannan, Perth, and Stirling meet and the south-east corner of the shire near the Black Loch. This line is 21 miles in length. The total area of the shire (exclusive of inland water, tidal water, and foreshore) is 288,842 acres, or 451·3 square miles. It is thus the 20th county for size in Scotland. In 1920–1 the county valuation was £1,192,161.

The shape of Stirlingshire is very irregular. Its main portion may be described as a roughly quadrangular figure with its greatest length lying east and west, the four angles being approximately defined by Stirling, Castlecary, Bardowie Loch, and Buchlyvie. To this central portion are appended a long triangular extension skirting the east side of Loch Lomond to Glengyle; a south-eastern portion about Denny, Falkirk, Slamannan, and Grangemouth, which contains the great industrial district of the county; and a northern salient, roughly triangular, across the Forth at Bridge of Allan.

The extreme north-west corner of the shire lies at the summit of Beinn Ducteach (1750 ft.), near the sources of the Water of Gyle, which flows into Loch Katrine. Descending Glengyle, the boundary between Stirling and

Perth lies in Loch Katrine to a point east of Stronachlachar Hotel, where it turns inward to the south bank, crosses to Loch Arklet, and, after skirting the east end of that loch, strikes southward to the summit of Ben Uamhe (1962 ft.). From this hill the boundary line holds south-east across Ben Dubh (1675 ft.) and Mulan-ant-Sagairt (1398 ft.) to the Duchray Burn, which it follows down to a point just below Duchray Castle. Here the line cuts across irregularly in a south-eastern direction, round the east side of Drumore Wood and just west of Gartmore, and strikes the Kelty Water, which it follows, along the north edge of Flanders Moss, until its confluence with the Forth at Barbadoes. Henceforth the winding river forms the march eastward until just above Stirling. Here the Allan Water flows into the Forth from the north; and the boundary, crossing the Forth, runs up the Allan until within a mile from Dunblane, when it diverges to the north-east up the Wharry Burn. Passing through Glentye, the boundary then turns southward round the east flank of Loss Hill (1771 ft.), where Clackmannan succeeds Perth as the adjoining county. West of Colsnaur Hill (1832 ft.) the march descends the valley of the Menstrie Burn until opposite the village of that name. It then swerves westward towards Blairlogie, and, swinging south-east once more, follows the River Devon down to Bridge End. Striking thence west and then south, it joins the Forth opposite Lower Taylorton. From here to the mouth of the Avon on the south bank the estuary of the Forth is the boundary, Clackmannan and then Fife lying across the tidal water.

From Inveravon the march between Stirling and Linlithgow ascends the River Avon and the Drumtassie Burn to a point a little north of Westfield, where Lanark is touched. This shire continues to march with Stirling along the boundary which, making a sharp re-entrant round Hillhead, crosses the Black Loch and holds north-west until it descends again into the upper valley of the Avon, a little east of Fannyside Loch. Here an outlying portion of Dumbarton is touched. Beyond this another re-entrant occurs, the march line swinging north-east to Jawcraig and then to the west again by Arns and Glenhead to Castlecary. From this point westward the River Kelvin generally defines the boundary, but the line is sometimes north and sometimes south of the river in a curiously erratic fashion. A little west of Kirkintilloch (in Dumbarton) the northwest corner of Lanark touches our county, and forms its boundary along the Kelvin until a point south of Bardowie Loch. Henceforth the main portion of Dumbartonshire is skirted by the boundary, which runs irregularly in a northwestern direction by Mugdock and the Allander Water and Auldmurroch Burn, and so down Burn Crooks (the headstream of the Finnich). A little below the junction of the Carlingburn the boundary strikes westward, and, passing over Gallangad Muir, descends by Wester Cameron into the Catter Burn, which it follows to the Water of Endrick, south-east of Drymen. The Endrick then forms the boundary to its mouth in Loch Lomond.

In the southern end of the Loch the boundary between Stirling and Dumbarton forms a sinuous line. Striking westward from Endrickmouth, and passing between Torrinch and Inchcailloch, it swings round the west side of

Inchfad, and so between Inchmoan and Inchcruin and round the east side of Inchlonaig. Thus the islands of Clairinch, Inchcailloch, Inchfad, Inchcruin, Ceardach, and Bucinch are in Stirlingshire. After rounding Ross Point, the march pursues a central course up the Loch until opposite Ardvorlich Point on the Dumbarton side. Here the boundary holds to the east, and strikes the shore opposite Island-i-Vow. Climbing the steep slope eastward to the summit of Beinn-a'-Choin (2524 ft.), it swings to the north over Maol-an-Fhithich (2000 ft.) and Stob-nan-Eighrach (2011 ft.), and so completes the circuit at Beinn Ducteach.

4. Surface and General Features.

The surface of Stirlingshire may be divided into five well-marked areas, corresponding to the geological structure of the rocks beneath. The extreme north-western portion, beyond a line drawn from Duchray to Inchcailloch on Loch Lomond, belongs, geologically and topographically, to the Highlands of Scotland. It is rough moorland, partly wooded with natural birch and plantations of larch and fir, and rising into magnificent mountain scenery. The general contour of these hills is determined by the manner in which their constituents have yielded to decay. Though there is a great cliff at the summit of Ben Lomond, their shape is generally rounded, owing to the fine detritus into which the slaty rock breaks down, and to the very uniform rate of disintegration. In this respect the Stirling Highlands contrast sharply with the granitic and gneissose mountains further north, which weather along joint-planes into large quadrangular blocks, causing a skyline of "tors" or rugged bosses. The highest summit is Ben Lomond (3192 ft.), and within this area are the following other peaks over 2000 ft.: Beinn-a'-Choin (2524 ft.); Ptarmigan (2398 ft.); Maol Mor (2249 ft.); Stoban Fhainne (2144 ft.); Cruinn-a'-Bheinn (2077 ft.); and Stob-nan-Eighrach (2011 ft.). This line of heights east of Loch Lomond is part of the central watershed of Scotland, the streams flowing off it on the east towards the Forth, and on the west towards the Clyde. The Highland district also includes the eastern half of Loch Lomond, gemmed with wooded islands, the picturesque Loch Arklet, and part of Loch Katrine, at its northern end.

In sharp contrast to the Highland Area is the Lowland country which stretches eastward along the basin of the Forth, skirting the uplands of Kippen and the Fintry and Gargunnock Hills, and expanding at the lower end into the broad low-lying country at Stirling, St Ninian's, Airth, Dunipace, Denny, Larbert, Falkirk, Bothkennar and Polmont. Apart from the flat carselands, this region is undulating in surface and is rich arable country, finely wooded. It is continued on the south side of the Campsie Fells and Kilsyth Hills along the valleys of the Bonny and the Kelvin. From the western end of the carseland the low country sweeps west and south by Buchlyvie and Balfron into the valley of the Endrick, descending on Loch Lomond. That portion of the Lowland which lies in the Forth valley above Stirling is purely agricultural in character; but the eastern portion, around Bannockburn, Denny, Camelon, Larbert, and Falkirk, contains the great mineral fields which give the shire its economic importance. Two products of its

Carboniferous strata, coal and iron, have made Eastern Stirlingshire one of the industrial centres of Scotland.

The middle portion of the shire is occupied by a volcanic *massif*, known in its different parts as the Campsie Fells, Kilsyth Hills, Fintry Hills, Gargunnock Hills, and Touch Hills. These names are very loosely applied. Collectively they are termed the Lennox Hills. A portion of the adjoining volcanic *massif* known as the Kilpatrick Hills, lying between Loch Lomond and the Blane valley, is included in the county. The Campsie Fells stretch eastward from the Blane valley. At the west end they throw off the Strathblane Hills, and southward they are bounded by the valleys of the Glazert and the Kelvin. To the east the Campsies are continued by the Kilsyth Hills, which slope down upon the low ground at Dunipace and Denny. This long ridge of elevated ground is grass-covered, and reaches its highest altitude in Earl's Seat (1894 ft.), the summit of the Campsies. The highest point of the Kilsyth Hills is Laird's Hill (1393 ft.), overlooking the town of Kilsyth. Northward the Campsie Fells descend into the green valley of the Endrick, around which are the Fintry Hills, soft and verdant mostly, abrupt in parts. Their northern summit is Stronend (1676 ft.). East and north the Fintry Hills are prolonged by the Gargunnock Hills (highest point, 1591 ft.), which overlook the valley of the Forth about Kippen and Gargunnock, and by the Touch Hills to the south of Stirling. The grassy surface of all these hills affords splendid sheep pasture. An extension of the Gargunnock Hills runs south-eastward through Earl's Hill (1443 ft.) and Dundaff Hill (1157 ft.), and looks across the Upper Carron valley towards the eastern end of the Kilsyth range. Further to the east the upland country descends gently through St Ninian's parish, and sends out its last spur well into the Lowlands at the Torwood. These Carboniferous lavas of Central Stirlingshire, owing to their horizontal outcrop and the interstratified beds of softer tuff, weather into terraced escarpments which form a striking feature of their scenery. The average height of the volcanic plateau is about 1250 feet.

In the south-eastern portion of the county, in the parishes of Falkirk, Slamannan, and Muiravonside, occurs another area of high ground, all over the 500 foot level, the surface of which is composed largely of sandy soil overlaid by wet moss. The progress of cultivation and plantations of trees have only partly availed to remove the characteristic bleakness of this upland area.

Lastly we have the district north of the Forth, extending from the fertile carseland up into the Ochils, and reaching at the north-west corner of the shire a height of 1771 ft. This area includes the picturesque, isolated summit of Dumyat (1375 ft.), rising almost perpendicular in terraced cliffs from the carse, and commanding a magnificent view over the Forth estuary and eastern Stirlingshire. Between Dumyat and the main mass of the Ochils the burn of Menstrie flows down in a deep, rich little valley to join the Devon in Clackmannanshire.

5. Watershed, Rivers, Lakes.

The river system of Stirlingshire is conditioned by the central *massif* of the Lennox Hills. Along the north front of these hills flows the River Forth, the estuary of which, below

Stirling, swings round to the east and south. Numerous small streams, of which the Boquhan and the Bannock Burn are most important, flow down from the Gargunnock Hills into the Forth. The south side of the Lennox Hills is defined by the valleys of two streams flowing in opposite directions. Rising north-east of Kilsyth, the River Kelvin flows south-westward for a distance of some 22 miles until it falls into the Clyde at Partick, the western suburb of Glasgow. Up to the neighbourhood of Bardowie Loch, a distance of about 12 miles, the Kelvin generally forms the boundary of the shire. The Garrel Burn, which it receives at Kilsyth, descends in a series of waterfalls traversing a deep gorge, in the rocky sides of which coal seams may be observed amid the sandstone. Eastward from Kelvinhead and Castlecary the Bonny Water, rising in the Dumbarton parish of Cumbernauld, flows in the opposite direction to join the Carron below Dunipace.

The central upland district, which divides the Forth valley from the Kelvin-Bonny Water valley, is itself sub-divided by the valleys of two considerable streams flowing respectively east and west. The River Carron, wholly a Stirling stream, rises in Carron Bog, in the parish of Fintry, about 1000 feet above sea-level, and has a course of some 25 miles eastward until it enters the Firth of Forth at Grangemouth, where it receives on its right bank the Forth and Clyde Canal. The upland parts of the Carron valley, where it descends among the Lennox Hills, are wild and Highland in aspect. Issuing from the hills in the ravine of Carron Glen, and passing over the turbulent Linnspout of Auchenlilly, it winds sedately through the carseland past

Dunipace, Denny, and Larbert, presenting occasionally
scenes of gentle lowland beauty amid a valley greatly marred
by industrialism, but formerly famed for "the bonny banks
of Carron Water." The mouth has been straightened to
assist navigation connected with the port at Grangemouth.

In Carron Bog also rise some of the headsprings of the
Water of Endrick, the main stream of which descends from
the north between the Fintry and the Gargunnock Hills.
At Todholes the stream makes a sharp turn to the west,
and flows in a winding course of some 29 miles along the
north side of the Campsie Fells, past Fintry and Balfron
and Drymen until it enters Loch Lomond west of Buchanan
Castle. From Drymen station to Loch Lomond the Endrick
Water forms the boundary between Stirling and Dumbarton. In its course the Endrick receives a number of tributaries, of which the most important is the Blane, or Ballagan.
Rising in Earl's Seat (1894 ft.) it flows first southward in
a romantic glen, forming at the Spout of Ballagan three
beautiful cascades with a descent of 70 feet. Turning westward, the Blane flows past Strathblane village and through
the fertile valley of Strathblane, to join the Endrick at
Killearn, after a course of some $9\frac{1}{2}$ miles. The Dualt Burn,
which the Blane receives shortly before joining the Endrick,
has a beautiful cascade of 60 feet in a wooded glen. Near
its mouth the Water of Endrick receives from the north
Mar Burn, which rises in Bheinn Bhreac (1925 ft.).

Strathendrick—"sweet Innerdale"—contains scenery of
great beauty. One of the best known spots in the valley is
the Loup of Fintry, where the river leaps a precipice more
than 90 feet in height.

Sending the Carron towards the Forth, and other streams to join the Endrick which flows into the Clyde, the high central plateau about Carron Bog may be regarded as the watershed of Stirlingshire. At one point, indeed, the infant waters of the Endrick and the Carron are less than a mile apart, being separated by Dungoil (1396 ft.) and Gartcarron (1006 ft.), the two northernmost outspurs, towards Fintry, of the Campsie Fells.

The River Avon, which rises in the detached portion of Dumbarton, and has a total length of 21 miles, enters Stirlingshire west of Slamannan, and, traversing the south-eastern corner of the county, forms the march with Lin-lithgow from the point where it receives the Drumtassie Burn to the Firth of Forth at Inveravon.

The River Forth is formed by the junction of the Duch-ray Water and the Avondhu, both rising on the eastern flanks of Loch Lomond. For nearly 7 miles the Duchray Water divides Stirling and Perth: the Avondhu is wholly a Perthshire stream. The Forth itself becomes the county boundary at the confluence of the Kelty in Flanders Moss, and remains the boundary for the rest of its course, with the exception of the salient on the north bank at Bridge of Allan. After winding in a magnificent series of loops—the "links of Forth"—across the broad central plain of Scotland, the river enters a more contracted reach of its valley between Stirling and Bridge of Allan. Here the valley is a line of fracture caused by the great fault which truncates the volcanic area of the Ochils. It has been widened by erosion which has left the hard igneous cores of Stirling rock and Abbey Craig facing each other across the narrow valley.

The carse is here reduced to less than a mile in width, and the river winds across it in noble loops. The estuary proper may be said to commence about Airth, although the tide is felt certainly as far up as the cruives at Craigforth Mill, three miles above Stirling, where the rise and fall is at least 10 feet in vertical height. Beyond this point the influence of the tide is checked by a bar of Old Red Sandstone which crosses the river at this spot. Above it the river is navigable by rowing boats as far west as Flanders Moss.

In the upper part of its course, amid the Highland area, the descent of the headwaters of the Forth is extremely rapid, but thereafter the fall is very gradual and the current slow and tranquil. The Duchray Water, which is only $13\frac{3}{4}$ miles in length, rises at a height of 3000 feet upon Ben Lomond, yet its confluence with the Avondhu is only 80 feet above sea-level. Along the northern boundary of Stirlingshire, from the junction of the Kelty to Inveravon, the length of the Forth is about 53 miles. Opposite Alloa the river loses its winding character and begins to expand into the estuary. It is here about 400 yards in width; at Airth it attains a width of half a mile, and at Inveravon the estuary is a broad saltwater firth over 3 miles in width. The total length of the river and its estuary, from the source of the Duchray to the North Sea, may be computed at $116\frac{1}{2}$ miles.

Among the Stirlingshire tributaries of the Forth need only be mentioned the Boquhan Burn and the Bannock Burn. Rising east of Stronend (1676 ft.), the highest summit of the Fintry Hills, the Boquhan Burn flows through a little glen of singular beauty and joins the Forth at Bridge of Frew. The historic Bannock Burn rises north of Earl's Hill (1443 ft.) in the south-eastward expansion of the Gargunnock Hills, and flows north by east, with a course of some 14 miles, to the Forth below Stirling. Near the village of Bannockburn the stream flows between steep banks, above which lay Milton Bog and Halliert Bog, now drained; and lower down, its course in the 14th century lay through a morass intersected by ditches and lagoons. These natural features of the Bannock Burn played an important part in the two days' fighting.

A considerable part of Loch Lomond, on the eastern side from Endrickmouth to a point about 2 miles north of Inversnaid, lies within the county. Loch Lomond is the largest lake in Scotland, being 22 miles in length and 5 in greatest breadth at its south end, with an area of 27 square miles. The surface of the water is 23 feet above sea-level. It is drained by the River Leven, which, emerging from the Loch of Balloch, flows into the Clyde at Dumbarton. The deepest part of the Loch (105 fathoms) is in Stirlingshire off Culness, south of Inversnaid. It contains abundant salmon, sea trout, lake trout, pike, and perch, and is celebrated for the Loch Lomond herring or "powan," a fresh water herring of great delicacy. The fishing is entirely free. On the Stirling side, near the southern end, the scenery of Loch Lomond is soft and verdant, embosomed in the rich plain of Drymen and Killearn, and studded with romantic wooded islands, while across the water the stern hills behind Luss lend character to the scene. Further north along the Stirling shore succeed the rugged mountains, but these

Loch Lomond

never reach the savageness of some parts of the Scottish Highlands. Their outlines, though imposing through their mass, are rounded, and their flanks are richly wooded. The lake is a true rock basin, having been hollowed out by the action of ice. The diversity and contrast of its scenery is due to the fact that it occupies a "transverse" valley, cutting across the "strike," or general direction of the outcropping beds of rock. Loch Lomond has been truly termed the Queen of Scottish lakes. In the shores of the Loch about Inversnaid, Stirling possesses a district rendered classic in English literature by its association with Rob Roy. Craig-royston, or Rob Roy's Cave, is about a mile north of Inversnaid.

A small portion of Loch Katrine, from its upper end in Glengyle about $2\frac{1}{2}$ miles along the right bank to a point opposite Stronachlachar, is in Stirlingshire. It is 364 feet above sea-level. The head of the Loch lies amidst stern desolate scenery, and seemed to Dorothy Wordsworth in 1804, "like a barren Ullswater—Ullswater dismantled of its grandeur, and cropped of its lesser beauties." Modern plantations have little altered the austerity of the scene, and the view from Stronachlachar Hotel is one of the wildest in Scotland. Loch Katrine, like Loch Lomond, derives its abrupt scenery from the fact that it lies in a transverse valley. Since 1859 the Loch has been utilised for the water supply of Glasgow, and the operations connected therewith have considerably raised its level. Leaving the Loch at a point in Perthshire, about 2 miles east of Stronachlachar, the two aqueducts enter Stirling in the valley of the Duch-ray, south-west of Loch Ard. Passing by varying courses along the eastern face of the Ben Lomond group, they descend into the lowland country near Balfron Station, skirt the Campsies to Strathblane, and pass out from the county near Mugdock. At this point there are two large reservoirs: Mugdock, containing 548 million gallons, and Craigmaddie, containing 702 million gallons. At the normal consumpt of a busy season, this would provide storage supply in reserve for about a fortnight. The water is so pure that it undergoes no filtration, but it is strained through fine copper sieves in wells at the outlet from each of the reservoirs before it is admitted to the mains leading to the city.

Between Loch Katrine and Loch Lomond lies Loch Arklet, 455 feet above sea-level. It is wholly within Stirling, of which its eastern end forms part of the boundary. This lonely little lake was originally somewhat over a mile in length and a quarter of a mile in breadth. It drains into Loch Lomond by the Arklet Water, which near its mouth at Inversnaid makes a beautiful little fall of 30 feet. At its west end is Corrie Arklet, the residence of Rob Roy and birthplace of his wife Helen Macgregor. In Glen Arklet is the site of the fort erected in 1713 to curb the Macgregors. For a short time it was commanded by General Wolfe, then an officer of the Buffs. Loch Arklet, like Loch Katrine, has been utilised for the water supply of Glasgow. It has been connected by tunnel with Loch Katrine, and has been raised above its original level by 22 feet, producing a sheet of water $2\frac{1}{8}$ miles in length. Its available storage capacity is 2702 million gallons. The storage capacity of Loch Katrine at its raised level under the Act of 1885 is 9853 million gallons. An Act was obtained in 1919 to raise the level of Loch Katrine still further, and work for this purpose is now progressing. The effect will be to add 4349 million gallons additional storage to the combined capacity of Loch Katrine and Loch Arklet, bringing the total storage in the two lochs up to 16,914 million gallons.

In the lowland district are several small lochs, of which need be mentioned only Loch Coulter, on the eastern slopes of the Touch Hills towards Dunipace; Black Loch, in Slamannan parish, but partly within Lanark; and Loch Laggan, south-west of Kippen. Black Loch is used to fill the Hillhead Reservoir for the Monkland Canal, and Loch Coulter has lately been adapted to form a distributing reservoir for Eastern and Central Stirlingshire.

6. Geology and Soil.

In studying the rocks which compose the earth's solid crust, geologists find it convenient to classify these by their mode of origin. From this view point, rocks may be assembled in three great groups. First are the Igneous Rocks, which have cooled from the molten state. Such are granite, diorite, basalt, and gabbro. The igneous rocks are found as intrusive masses breaking through other rocks. Where their shape is irregular, such intrusive igneous masses are known as "bosses"; where they occur as horizontal sheets, they are called "sills"; when the sheet breaks vertically through the surrounding rocks, the name "dyke" is applied. If an intrusive igneous mass reaches the surface, volcanic action ensues, producing lava flows and "tuffs" or ash-beds. Such deposits form a branch of the igneous group and are termed Eruptive Rocks.

The second group of rocks are called Aqueous or Sedimentary. Four great destructive agencies—frost, heat, water, wind—are constantly breaking rocks into boulders, pebbles, and finally sand and earth. So the blanket of soil which veils the solid bedrock has been formed. The ultimate destination of all such disintegrated material is the sea, into which it is carried by rivers. Every land surface is thus an area of waste or "denudation," and every sea basin is an area of deposition. The vast masses of sand, clay, and mud brought down by rivers slowly accumulate beneath the oceans, or in the basins of fresh water lakes, as sheets of sedimentary material, which in course of ages may become consolidated and uplifted by crustal movements to form aqueous rocks. In this way sandstone, shale, and clay have originated. Aqueous rocks occur in "strata" or layers, owing to their mode of formation as sheets of land-waste accumulated and sorted out in oceanic basins. Originally these strata or beds were horizontal, but often they have been thrown into folds and sometimes fractured by crustal movements. Fractures may be accompanied by slipping down of the beds along one side of the break. In this case they are known as "faults."

Aqueous rocks often contain fossils. These are remains of animals and plants that flourished in the sea where the rock was formed, or were washed into it by rivers from the adjoining land. One or two aqueous rocks consist wholly of such organic remains. Much chalk is formed of the minute shells of animals; coal is composed of the trunks and leaves of ancient trees. Such rocks are classified in a sub-division of the aqueous group as Organic Rocks.

The third great group of rocks are called Metamorphic. These were originally either igneous or aqueous, but have since been exposed to intense crustal pressure, heat, and chemical action. Thus their structure and composition are radically altered. Sandstones are transformed into quartz-ites, clays and shales into slate or schist, limestones into marble, granite into gneiss.

In the geological structure of central Scotland the cardinal feature is the Highland Boundary Fault, which stretches from Stonehaven in a south-westerly direction to the mouth of the Clyde near Helensburgh. South of this fracture the rocks have slipped many thousand feet, so that stratified formations of more recent date have been thrown down against metamorphic rocks of vast antiquity, exposed by denudation of the aqueous deposits that once covered them. A fundamental division is thus created by the fault. North of it the country consists mainly of hard metamorphic rocks which have offered stubborn resistance to erosive agencies, so that they stand up as the rugged Highland mountain masses. South of the fault the softer sediments of the Central Plain still remain, preserved by the down-throw of the great fault. This geological distinction between Highlands and Lowlands lies at the root of the political, social, and economic development of Scotland. In the High-lands the hard metamorphic rocks, deficient in mineral resources, weathering into rugged mountains with poor and sandy soils, have made this country the refuge of the ancient Celtic race, driven out of the rich Lowlands into a barren territory where their only occupations were war and the chase. On the other hand, the fertile Lowlands, rich in minerals, have become under the intruding Saxon the hub of Scotland's prosperity and the core of her national life.

By observing their field relationships and fossil contents, geologists have arranged the aqueous rocks in three groups, Primary, Secondary, Tertiary, in order of age. Only the Primary group is represented in Stirling. It is subdivided into the following systems, the lowest or oldest being given last:

> 6. Permian.
> 5. Carboniferous.
> 4. Old Red Sandstone.
> 3. Silurian.
> 2. Ordovician.
> 1. Cambrian.

The sedimentary rocks of Stirling belong almost entirely to the Old Red Sandstone and Carboniferous systems.

The Highland Fault enters Stirling south-west of Aberfoyle, and traverses it to Loch Lomond, which it crosses by the islands of Inchcailloch, Creinch, and Inch Murrin. The north-west corner of the county thus belongs to the Highland area, and forms the mountain mass of Ben Lomond. Slate to the north of the fault is succeeded by coarse quartzites or schistose grits which in their turn give place to the mica-schist forming the Ben. These rocks are intensely folded, causing much repetition of the beds. Immediately north of the fault, wedged in between the metamorphic rocks and the down-thrown edge of the Old

Red Sandstone, occurs a zone of slates, cherts, and grits, known as the Highland Border Rocks. They seem to belong to the Cambrian and Ordovician periods.

To the south of the fault a broad band of deposits belonging to the Old Red Sandstone crosses the county, from Bridge of Allan to the Finnich Burn, where it passes beyond the county boundary. They consist of coarse conglomerates overlaid by red and grey sandstones, and have been thrown into a great "syncline" or trough-like fold. Where they meet the Highland rocks at the great fault, the Old Red Sandstone strata forming the north-western edge of the trough are tilted into a vertical position. Along the south-eastern edge of the trough the Old Red beds pass under the Carboniferous. A sharp unconformity occurs between these Old Red Sandstone rocks and the overlying series of deposits, which belong to the Carboniferous series. The Carboniferous rocks dip to the south-east and rest on the denuded edges of the Old Red rocks, which dip north-westward to the axis of the syncline. The older rocks must therefore have been first thrown into their synclinal posture, and then extensively denuded before the Carboniferous rocks were laid down on the worn upturned edges. A prolonged gap is thus indicated in the geological history of the shire. Between Balfron and Kippen a local fault, running E.N.E. to W.S.W., has let down the Carboniferous rocks into contact with the Old Red rocks, obliterating the unconformity.

The Carboniferous rocks fall into three groups: Lower, Middle, and Upper. The Lower group, forming part of the Calciferous Sandstone series, consists of red sandstones and blue, red, and grey clays, with cement-nodules and marls. They occupy the area between the Old Red rocks and the Campsie Fells and Kilsyth and Gargunnock Hills. They also appear to the south of these hills, and are well exposed in the Ballagan Burn near Strathblane. The high ground of these ranges consists of a vast outpouring of volcanic rocks, lavas and tuffs, which forms a remarkable feature of the Lower Carboniferous system in Scotland, and is continued south-westward through the Kilpatrick Hills and the uplands of Renfrew and northern Ayr. In Stirling these lava flows reach a thickness of 1000 feet, and consist mainly of porphyritic dolerite or andesite. Some of the eruptive vents may still be traced. One was at Meikle Bin, and another at Dumgoyn. Along the southern slopes of the Campsies the successive lava flows may be traced mile after mile as a most striking series of parallel rocky escarpments. Eastward, between Stirling and the Carron Water, the volcanic rocks thin out and pass into the blue shales and cement stones which are the normal Lower Carboniferous deposits.

South of the Campsie Fells and Kilsyth Hills follow the Middle Carboniferous rocks. These are divided from the volcanic plateau by a fault which stretches from Strathblane eastward towards the Carron Water. They form a band of irregular width extending from the Forth below Stirling by Denny, Kilsyth, and Campsie to Mugdock. Two zones of limestones are distinguishable, separated by a coal-bearing group with clay-ironstones, forming the Kilsyth mining area. In the lower zone occur the Index, Calmy, and Castlecary limestones, in the upper zone the

Hosie and Hurlet limestones, with a bed of alum shale below the latter. Near Kilsyth the beds are intensely folded, a remarkable instance being the "anticline" or archfold known as "The Riggin."

Lastly, in the eastern corner of the shire, occur the rocks of the youngest or Upper division of the Carboniferous period, consisting of the Millstone Grit and the Coal Measures. Their lowest beds form the Millstone Grit, consisting of sandstones and fireclays, with coal and ironstone. They are found in an irregular area stretching from Carnock to the shire boundary south-east of Castlecary. Another band extends from Polmont south into Linlithgow, and there is a third patch north of Kirkintilloch. The Millstone Grit yields a poor cold soil, and is locally known as "Moorstone." The remainder of the shire is occupied by the true Coal Measures, the highest and most recent of the Carboniferous deposits, reaching their maximum development at Falkirk, Stenhousemuir, Larbert, Slamannan and Grangemouth. The chief seams are the Coxhead, Craw, Splint, and Virtuewell, the latter being highest. The Banknock or Dennyloanhead Coalfield is an outlier from the Coal Measures, faulted down between the Carboniferous Limestone and Millstone Grit.

The greater portion of the shire north of the Forth at Bridge of Allan belongs to the large volcanic mass of the Ochils, dating from Old Red Sandstone times. The hills are built up of sheets of lavas and conglomeratic tuffs showing a terraced structure like that of the Campsies. This phenomenon is strikingly displayed upon Dumyat. The volcanic area is truncated to the south by a fault letting down against it the Carboniferous strata which form the contracted valley of the Forth.

At various points the sedimentary rocks have been breached by intrusive igneous masses. One of the largest is the great sheet of diabase which forms the rock of Stirling Castle and Abbey Craig. It is continued in broken masses, doubtless connected underneath, through Dunipace and Kilsyth to Mugdock. The white sandstones and shales among which this sheet is intruded have been much altered by contact with the molten mass. This is particularly evident at the Castle rock, where the shales are baked into a compact spotted slate and the sandstones into a quartzite. Numerous basalt or dolerite dykes traverse the shire from east to west. These are probably of Carboniferous age. There is a fine basalt boss, beautifully columnar, about a mile south of Strathblane. Much use is made of these basaltic intrusions for road metal.

No sedimentary rock younger than the Carboniferous period exists in Stirling. Everywhere these ancient formations are overlaid by the superficial deposits of the Ice Age, which intervened between the Tertiary period and that in which we live. Most important of these is the Boulder Clay, a stiff, tenacious, earthy clay full of boulders, which are often smoothed, polished, and scratched by ice-action. The Boulder Clay is detritus scoured from the land surface by great glaciers which then gripped the country in their icy fingers. Other relics of the Ice Age are the ridges and hummocks of gravel called "Kames." These were formed by torrential sheets of water when the great ice masses were melting. A fine example may be seen from the railway west of Polmont Junction. The Ice Age has also left its mark on the bare upland rock surfaces, which are smoothed and scored. These markings show that the ice-sheet moved south-eastward from the Highland plateau into the basin of the Forth. Ice markings are traceable on Ben Lomond up to 2250 feet. As Loch Lomond is more than 600 feet in depth, the ice lying in this basin must therefore have been nearly 3000 feet thick.

Another topographical feature due to the Ice Age is "Crag and Tail" structure. Where a hard intrusive igneous mass traverses softer sediments, the erosive force of the moving ice has been broken, and the rocks on the leeward side are preserved from destruction. A sheltered ridge is thus formed behind the intrusive mass, and on this ridge drift accumulates to form a "tail." The rock of Stirling Castle, and the long ridge on which the old town huddles, make an excellent instance of this phenomenon. Abbey Craig is another example.

Along both sides of the Forth estuary raised beaches may be seen at 50 and 100 feet above the present sea-level. These are flat terraces of clay, gravel, or sand, with marine fossils, and indicate great changes in the distribution of land and sea since the glacial period. After the ice retired, submergence followed, at the close of which the land seems to have risen and the 100 foot beach was cut out. The elevation probably continued until Britain was joined to the Continent, when the present fauna and flora were introduced. Submergence again took place, and the 50 foot beach was formed. Lastly, a fresh upheaval resulted in the present coastline.

Among the geological curiosities of the shire are the Whangie, the Auld Wives' Lifts, and the Mounds of Dunipace. The Whangie, on Auchineden Hill, is a huge fissure in one of the plugs of lava which fill the old volcanic vents in the Kilpatrick Hills. It is about 30 feet deep, 4 feet in average width, and nearly 20 feet in length. The Auld Wives' Lifts on Craigmaddie Muir consist of two

The Auld Wives' Lifts, from the west

immense blocks of sandstone surmounted by a third, the whole being nearly 13 feet in height. It has often been described as a dolmen, but is purely natural, though primeval man dressed the capstone and incised a circle on it. A human origin has also been claimed for the Mounds of Dunipace, but these are residual masses of the 100 foot beach left in the loops of the ancient course of the Carron, which has carried away the rest of the littoral deposits.

The finest soils in Stirlingshire are produced by the carselands bounding the Forth. "A loop of the Forth," says the old rhyme, "is worth an earldom in the North." The carse-clay, forming the 50 foot beach, extends about 30 miles along the river from Bo'ness to Gartmore, and varies from 1 to 4 miles in breadth. Immediately under the surface is a stiff yellow clay much used for bricks, below which is a blue clay, very wet. Stones are comparatively rare, but beds of marine shells occur. Originally much of the carse-land was covered with peat, and this explains the lack of villages upon it, also the disposition of the old roads, which avoided the mosses. Thorough draining, subsoil ploughing, and liming have now made the carse the best farming land in the shire, particularly in the rich clays along the river, where freshwater alluvium is mingled with the old marine deposits. It is especially good for wheat and beans. At the edges of the carse, the soil, being mixed with sands and gravels washed out of the 100 foot beach, is of a poorer quality.

The deposits of the 100 foot beach yield the soil known as "dryfield." The strata are mainly sand and gravel, and form a soil poorer and lighter than that of the carse. But sometimes, for example in Strathendrick, the dryfield attains a high degree of fertility. Potatoes and turnips are chiefly grown on this soil. A colder and poorer soil is found over the Campsie Fells and the Kilsyth Hills, where there is much boulder clay. These rising grounds form excellent pasture, and there is good grazing over much of the Highland area, except at the higher levels, which are chiefly moorland. Peat mosses still remain in places, for example over the Carse at Offrance and the Millstone Grit in Slamannan. Formerly they were much more extensive.

7. Natural History.

The flora and fauna of Stirlingshire exhibit no very special characteristics marking the country out from the rest of Scotland. Geological research has shown that the plants and animals at present inhabiting Britain reached it in a far distant age when this country was still united to the Continent. But before the entire European fauna and flora had made their way into Britain, the connexion with the Continent was severed by the formation of the North Sea and English Channel. Hence arises the comparative poverty of animal species in Britain, which has only about 40 different land mammals, whereas Germany has about 90. The same remark is true of the flora, particularly in regard to trees. Most of the trees now to be seen in Scotland have been introduced in comparatively recent times by human agency. Thus the only important native conifer is the Scots Pine, although other trees of this family, such as the European larch, the Douglas fir, and the Norwegian spruce, have been introduced and now flourish in many parts of the country. But it is probable that other agencies besides the early separation of Britain from the Continent have played their part in producing Scotland's poverty in native trees. As the ice-sheet vanished the ground was occupied by an arctic-alpine type of vegetation, with such shrubs as alpine birches and willows and the creeping azalea, which gradually gave place to forests of birch and pine. These in their turn were engulfed by a wet moorland type of vegetation indicating the prevalence of moister climatic conditions. Evidence is found in various places in Scotland in the succession of plants in the peat mosses that the moor was again occupied by pine forest, which once again gave place to the moor. In regions of dry moorland with thin peat the trees have probably disappeared in comparatively recent times, largely through human agency. Natural wood still exists in some parts of Stirlingshire, for example in the Ben Lomond district, in the Torwood, and on the Touch Hills.

Botanists have divided Scotland into four zones of vegetation, defined broadly by altitude, though of course much modified by special conditions such as climate and soil. Along the shores there is a littoral zone, with the flora common to salt-marshes, links, and sand-dunes. This zone is not developed to any extent in Stirlingshire. The shore conditions are not really maritime, and many characteristic littoral plants are unrepresented. Marine plants occur as far up the Forth as Old Polmaise, about 3 miles below Stirling.

Next comes the zone of cultivation, extending roughly to a height of 1000 feet, which in Stirlingshire may be divided into two sections, each with a distinctive flora. At the lower levels is the carse-land of the 50 foot raised beach (p. 31), with a heavy clay soil, exceedingly fertile, and well suited for growing wheat. Its wild flowers are a well-marked group of the weeds of cultivation. These do not usually show the effects of soil, but are determined more by the system of cultivation. On the carse-land is also found a stray flora representative of pre-existing marshes. At a higher level are the sands and gravels of the 100 foot raised beach and the glacial deposits. The soil here is known as "dryfield" (p. 31), and being light is well adapted for the cultivation of potatoes and oats. The native plants belong to the characteristic type of Lowland Scotland. This zone represents the area anciently covered with forest, and is also the zone in which man has wrought the greatest changes in the flora. In this zone also are found the native broad-leaved trees. Of these the oak was originally the most important, but it has largely been ousted by the beech, which is not native to Scotland. The eastern lowlands of Stirlingshire, the valleys of the Forth and Kelvin, the skirts of the central volcanic *massif*, and the flat country about Drymen, belong to the zone of cultivation. The foothills within this area, formerly clad with forest, comprise some of the first reclaimed land in Stirlingshire. Throughout this zone plantations are numerous, and many fine specimens of trees may be found. Along the Forth the lines of pollard willows give a curious and distinctive character to the landscape. There is some native birch and oak copse on the banks of Ben Lomond and in the lateral glens.

Above this comes the sub-alpine zone, now the area of sheep farms, marked by the presence of moorland and peat bog, and rising to about 2000 feet. This zone also contains scattered wood, of which the Scots pine, mountain ash, and birch are the only native trees. The Scots pine is not usually found above 1500 feet. Along the streams grows the alder, and on the moorland wastes are found the bracken and many species of grass. Bracken is a feature of the Stirlingshire hills. It follows the deeper soils which are more or less moist, especially in winter, and hence forms a hill-foot zone or occurs in hill-flush patches. Where the soil passes into peat the grass yields place to heather, which can thrive in poorer soils. Among the heather grow the two common heaths, crowberry, blaeberry, and cranberry. In the wetter portions of the moor are to be found pink bell-heath, bog myrtle or gale, grass of Parnassus, lesser spear-wort, sundew, butterwort, bog-asphodel, cotton-grass, and many species of sedge and rush. Where the soil is very wet, bog moss grows in abundance, forming great areas of bog which are a constant feature in our Scottish moors. This sub-alpine zone reaches a great development in Stirlingshire, including the whole central uplands of the country. The lower hills are green, covered with grasses and bracken, including a certain amount of stream alluvium with grass-land and rushes, among which is the bog myrtle; the upper heather-clad hills are dark in hue.

Highest of all is the alpine zone. This is represented in Stirlingshire by the Ben Lomond district. Its flora is of special interest to the botanist, including many arctic plants which to-day form the last lingering representatives of our British flora in the Ice Age. As the climate grew warmer these plants were pushed upwards, and now find their last citadel of refuge on the high mountain tops, and sometimes on the coast. In this zone trees are few and stunted, but the dwarf willow and the birch are found at a very high level. Here also the blaeberry replaces the heather. Although the alpine zone in Stirlingshire is well developed in the Ben Lomond area, it is somewhat barren from a botanical view point, owing to the absence of the phyllites and limestones that support such a fertile alpine flora in the Perthshire Highlands. Among the alpine plants growing in the Ben Lomond district may be noted moss campion, alpine mouse-ear chickweed, rose-root, alpine lady's-mantle, mountain everlasting, dwarf chickweed, cranberry, crowberry, and alpine willow.

Rare plants are occasionally found in the neighbourhood of ancient inhabited sites, and are usually outcasts from gardens or otherwise attributable to human agency. For example, the deadly nightshade, formerly used for medicinal purposes, still grows on the Castle Rock of Stirling. Here also are found the herb Alexander, used for seasoning, the wild cabbage, lamb's lettuce, cat mint, and other interesting plants.

A census of plants in the shire has yielded 838 species and 80 varieties, including flowering plants, ferns, horsetails, and clubmosses. Of true mosses 238 species and 15 varieties have been recorded, including the rare *Tortula inermis*.

The fauna of the shire presents no features of special interest. Among the rarer mammals, the wild cat seems now extinct: its last recorded haunt was Strathblane. The badger is still occasionally found: an example from Leckie is preserved in the Smith Institute at Stirling. On the Forth the otter is found, and in its estuary the common seal may frequently be observed. The fox breeds freely in the Touch Hills and the Campsie Fells, and would soon be abundant were it not systematically kept down for the sake of the sheep stock. So early as 1283 Alexander III maintained a fox-hunter at Stirling. In a hard winter the red deer occasionally comes down from the Highlands and finds a temporary home in the Lowland area of the county. The

Blue or Mountain Hare
⅛ natural size

weasel, stoat, squirrel, common hare, and mountain hare are abundant. The mountain hare, found in the Ben Lomond district and the Gargunnock Hills, is one of Scotland's most interesting mammals. It does not occur in England, but is the only hare in Ireland. In Scotland the mountain hare has a special peculiarity in that it changes its colour from brown in summer to white in winter. The old black rat is probably now extinct, though a specimen was killed in 1886 at Kildean, and is now in the Smith Institute. In all 27 mammals are catalogued as occurring in the shire.

The number of birds recorded is 251. The golden eagle and peregrine falcon are found about Ben Lomond, along with the ptarmigan and the snow-bunting. The last bird is found also on the Touch Hills. Grouse (both black and red), capercailzie, pheasant and partridge are abundant. Along the Forth are found the dipper, kingfisher, water-rail, heron, mute swan, shelduck, wild duck, wigeon, teal, pochard, scaup, tufted duck, golden eye, goosander, red-throated diver, little grebe, cormorant, gannet, common gull, great blackbacked gull, and arctic skua. Loch Lomond also is rich in bird life, the species found there including the merganser and the great crested grebe. The Loch is also abundant in fish, containing salmon, pike, perch, flounder, trout, eel, and powan, the last having acquired considerable celebrity for its edible properties as the "Loch Lomond herring." In the River Forth salmon, trout, eel, and flounder are abundant, the salmon fishing being a local industry. Pike, perch, loach, and roach occur, but are some-what rare in the river. The sparling, a small fish of the salmon kind, visits the tidal reaches of the Forth in immense shoals twice a year.

The amphibians and reptiles include the frog, common toad, common and palmated newt, and the lizard, slow-worm, and adder. Insect life is abundant, but presents no features of special interest. Some 21 species of freshwater shells and 44 species of land shells have been collected in the shire.

Human interference with animal life manifests itself in two ways. On the one hand, man's action causes, directly

Red-breasted Merganser

or indirectly, the extinction of certain species; on the other hand it leads, directly or indirectly, to the introduction of new species. Both these phenomena may be illustrated from Stirlingshire. Thus in ancient days the wolf was a great pest in Scotland, and in 1283 Alexander III made allowance to his Treasurer for payment to "one hunter of wolves at Stirling." Early in the seventeenth century the wolf was still hunted there, but before the end of the century it had become extinct. In the same way the kite, which formerly existed in Stirlingshire, has now been exterminated from Scotland. The marten is a good example of an animal well on the road to extinction. Already rare in Stirlingshire by the end of the eighteenth century, it is now found only in the remoter parts of the kingdom. The same is true of the polecat, of which the last specimen was seen in Stirlingshire in 1879–80. Of animals introduced or preserved by man the fallow-deer is an example. Its attractive form led to its being valued as a denizen of the pleasure garden, and the Chamberlain's accounts for 1283 note allowances for the fallow-deer in the king's park at Stirling. Falcons were similarly cherished for sport, and Alexander III had a falconry at Dunipace. In the fifteenth century peregrines nested on the Abbey Craig and on the Ochils, and were carefully preserved. The ancient wild white cattle of Scotland, hunted by Bruce in the Torwood in the early fourteenth century, were in 1578 still preserved in the park of Stirling.

An instance of an animal introduced through direct human agency is supplied by the roedeer. In prehistoric times this creature ranged throughout Scotland, but destruction of the forests and increase of pasture brought about its practical extinction in the Lowlands by the end of the seventeenth century. The commencement of afforestation 100 years later led to its reappearance. Coming down from its Highland refuges into the new plantations, the roedeer appeared in Stirlingshire in the early nineteenth century and by 1845 had crossed the Pentlands. A similar story is that of the red squirrel. The dwindling forests had banished this graceful but destructive little creature from the Lowlands, when in 1772, at a time when much plantation was going on, a few were introduced into Stirlingshire, and are now (partly through subsequent introductions) common in all regions of Scotland. The American grey squirrel, introduced at Loch Long about 1890, reached the banks of Loch Lomond in 1906.

8. Climate.

The climate of Scotland is determined by a number of interacting factors, such as its position in regard to the great land and oceanic masses of the globe, its configuration and relief, and—most of all—its relation to the prevailing winds. Naturally it is intimately affected by the conditions which occur over the Atlantic, the great volume of water that washes its western seaboard. Over that ocean are found two more or less permanent pressure systems, which exercise a profound influence upon the winds, and hence upon the climate of Scotland. There is a low pressure system south of Iceland, and a high pressure system near the Azores. A third pressure system, influencing the British Islands, occurs over the European continent. It is high in winter, when the land surface is cold and there is much snow, but becomes low in summer as the land surface grows hotter. Over an area of low pressure strong winds tend to blow spirally inwards and upwards in a counter-clockwise direction; and as the wind rises it cools and its moisture condenses, so that low pressure areas are associated usually with mist and heavy rainfall. In a high pressure system, on the other hand, light winds blow spirally outwards in clockwise fashion, producing dry and fair weather.

During the winter season the climate of Scotland is affected mostly by the Icelandic and European pressure systems, between which a great swirling movement takes place, causing south-westerly winds to prevail, with much rain from the Atlantic. The general absence of extreme cold during winter is mainly due to the prevalent wind, which blows over Scotland from the warm southern regions of the Atlantic, and also causes a general drift towards our shores of the heated surface waters. In summer the Atlantic pressure system predominates, causing west or north-westerly winds laden with moisture gathered from the ocean. Thus the climate of western Scotland is purely oceanic in character, with mild winters and cool summers, being determined by the proximity of a great body of water with a fairly equable temperature. On an average the waters of the Atlantic are 3° warmer than the air. Upon the other hand, the mountain ranges of central Scotland prevent this oceanic climate from making its influence felt in the east. The plains fronting the North Sea have a climate which is more affected by the European pressure system, and therefore tends more to the continental type, with greater heat in summer and greater cold in winter. The waters of the North Sea are only 1° warmer than the air, and therefore its moderating influence upon the extremes of temperature is far less apparent. Although the mean annual temperatures of the east and west coasts are nearly equal, their mean summer and winter temperatures vary greatly, ranging from 39° and 56·8° F. in January and July at Portree to 37·5° and 59° F. at Perth.

From its geographical position, Stirlingshire is specially affected in regard to its weather conditions by the absence here of the antithesis between the western or oceanic and the eastern or continental types of climate. Between the Firths of Clyde and Forth the mountain barrier breaks down, and thus in this region the oceanic type of climate is carried much further east than elsewhere. Further north, the moisture laden winds from the Atlantic are intercepted by the Highland mountains, up which they pour, and by this cooling process most of their moisture is precipitated. But in the Central Lowlands the winds stream eastward still laden with moisture, and hence the whole of Stirling-shire is an area of comparatively heavy rainfall. The only portion of the shire which is approximately continental in its climate is the small strip of low ground between Airth and Grangemouth. The wettest region in the shire is found at its north-west corner, in the Ben Lomond district. Here the wet south-western winds, streaming up Loch Fyne, Loch Long and Loch Lomond, cool amid the sum-mits and their moisture condenses, producing a very heavy rainfall.

Owing to its position with reference to the Atlantic, Scotland generally is a country with a high mean rainfall. Only some 9 per cent. of its total area has a mean annual area of under 30 inches, while on the summit of Ben Nevis 160 inches of rain fall every year. In the Central

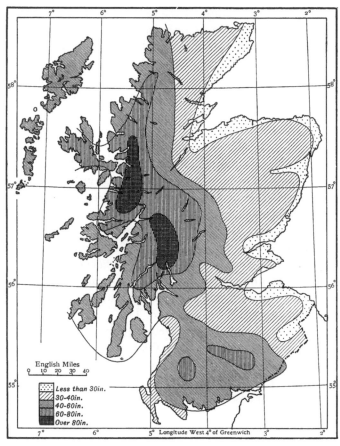

Rainfall Map of Scotland
(*after Dr H. R. Mill*)

Lowlands the annual rainfall measures 30 to 40 inches. Most of Stirlingshire lies within this area, and the yearly rainfall of the shire varies from about 35 inches in the eastern plain to about 115 inches in the north-west or Highland district. The mean annual temperature is 47·5° Fahrenheit. For January it is 38° F., and for July 59° F.

STATISTICS OF RAINFALL IN STIRLINGSHIRE
1920–4

(From *British Rainfall*)

Station	Height in ft. above sea	Inches of rain				
		1920	1921	1922	1923	1924
Muiravonside House ...	175	39·59	40·32	32·80	39·89	—
Polmont	320	37·87	36·81	27·90	34·89	32·29
Falkirk	107	39·24	35·32	28·64	41·81	37·83
Little Denny Reservoir	326	52·30	55·02	39·69	52·44	45·19
Kilsyth	707	69·87	66·19	50·59	63·95	58·13
Earlsburn Reservoir ...	1202	69·58	66·13	51·70	—	62·03
Touch House	35	57·70	51·93	—	—	—
Stirling	100	44·70	39·56	32·21	38·09	41·24
Gargunnock House ...	85	45·60	46·13	30·37	48·29	50·15
Kippen	150	53·05	47·62	—	54·80	51·91
Duchray Valley (Ben Lomond) ...	1800	96·00	96·40	81·55	110·55	91·80
Duchray Valley (Achaidh Burn) ...	1200	115·15	113·20	93·40	120·90	98·05
Duchray Valley (Head of main stream) ...	950	109·00	106·30	86·75	113·05	94·35
Stronachlachar ...	376	98·70	93·45	73·40	97·70	77·40

9. The People—Race, Language, Population.

The earliest inhabitants of Scotland appear to have been a wandering folk in a low state of culture, ignorant of husbandry, who eked out a precarious existence by hunting and fishing. They were doubtless descendants of the Old Stone Age folk of southern Britain, who followed the reindeer into Scotland as the climate of the Ice Age tempered and glacial conditions retreated northwards. To these primeval nomads may have belonged the hunters who stripped the stranded whales on the shores of the saltwater loch that, in those very ancient days, occupied the basin of the Forth to a point far west of Stirling. The "kitchen-middens," or refuse heaps of shells and bones split to extract the marrow, which are found along the margins of that vanished sea, are also attributed to these wandering folk (p. 71).

The next race to inhabit Scotland had developed considerable civilisation. They had abandoned the wandering existence of the Old Stone Age peoples for a settled agricultural life, and had tamed the dog, horse, ox, sheep, goat and pig. They had invented the arts of polishing stone, of weaving, and of making pottery. Above all, they had learned to commemorate their dead by heaping together huge cairns containing chambers in which the bodies were laid. Such chambered cairns served as family vaults through many generations. The dead were usually inhumed, and from their skeletal remains it is possible to infer the physical characteristics of this early race. They were a people of medium height, slightly built, with long head and oval face. Anthropologists believe that this race belonged to the dark-featured "Mediterranean" group of peoples of which the Basques are to-day a remnant. Their characteristic chambered cairns have not been observed in Stirlingshire, but implements and weapons of the Stone period are common, and prove that they once inhabited these parts.

Towards the end of the Stone Age, or perhaps about 2000 B.C., there appeared in Scotland the pioneers of the round-headed races who ultimately introduced the use of bronze. This people buried their dead singly in short cists of stone slabs, several examples of which have been found in Stirlingshire. It is in connexion with these short cist burials that we first come into actual contact with the bodily remains of the early races in Stirlingshire, and are enabled to form some idea of their physical character. The bones show a people of small stature and powerful physique, with short or round straight-back heads, and square determined jaws. A skull in excellent preservation, showing all these features, found in a short cist at Coneypark (p. 73) may be seen in the Smith Institute, Stirling.

With the introduction of bronze, probably about 1800 B.C., came the practice of cremation, so that we have no longer material on which to speculate about the races of this period. The later succession of peoples in Scotland is very obscure, but it seems probable that our country underwent slow changes in racial composition through the infiltration of various waves of Celtic peoples. The earlier Celts were known as Goidels or Gaels, and to these succeeded, probably about 500 B.C., the Brythons or Britons. The toponomy of Stirlingshire affords ample evidence of the presence of a Gaelic-speaking population in the past, and in the Highland area, at the extreme north-west corner of the county, place names are almost entirely Gaelic. Evidence for Brythonic names is less distinct. The introduction of a large English element into the population dates from the period subsequent to the eleventh century, when the Scottish monarchs encouraged Anglo-Norman immigration and re-fashioned the ancient Celtic institutions of their kingdom upon Teutonic lines. No doubt also a certain Norse element, in the days of the roving Vikings, found its way up the valley of the Forth. At present the names in the eastern plain of Stirlingshire are predominantly English, while in the central hill-country they are very mixed. English names do not seem to be found further up the Forth valley than Gartmore. In the census of 1911, 1742 individuals were returned as able to speak Gaelic as well as English, but of these only 164 were natives of the shire. In 1921 the Gaelic speaking persons were returned as 1277, of whom 5 spoke Gaelic only.

The population of the shire is very unevenly distributed, but bears a clear relationship to its geological structure. The densest population is crowded upon the industrial area over Carboniferous strata. A population of intermediate density, mainly agricultural, is found upon the area occupied by Old Red Sandstone; and in the Highland area of metamorphic rocks the population is exceedingly sparse.

In 1891 the population of Stirlingshire was 125,608; in 1901, 142,291, or 315 persons to the square mile—

an increase for that decade exceeded only by Linlithgow
and Lanark. In 1911 it was 160,991, of which 82,335
were males and 78,656 females. In 1921 it was 161,726
(an increase of 0·5 per cent.), of whom 81,424 were males
and 80,302 females. Of this population, in 1911, 81,384,
or 50·6 per cent., were enumerated within the burghs,
and 79,607, or 49·4 per cent., in the extra-burghal districts.
In 1921 the total burghal population amounted to 80,665,
or 0·9 per cent. less than in 1911, and the total extra-
burghal population numbered 81,061, or 1·8 per cent.
more than at the former census. Of the six burghs in the
county, intercensal increases in population were observed
in 1911 at Falkirk (population 35,574, increase 14·7 per
cent.), at Stirling (population 21,200, increase 15·2 per
cent.), at Grangemouth (population 10,219, increase 13·6
per cent.), and at Kilsyth (population 8106, increase 11·2
per cent.). At the burgh of Denny-and-Dunipace (population 5164) a decrease of 2·3 per cent. was noted, and
at the burgh of Bridge of Allan (population 3121) a decrease of 3·7 per cent. In 1921 the burghs of Bridge of
Allan (population 3579, increase 14·7 per cent.) and Stirling (population 21,345, increase 0·7 per cent.) showed
intercensal increases. Intercensal decreases were noted at
Falkirk (population 33,312, decrease 0·8 per cent.), Grangemouth (population 9699, decrease 5·1 per cent.), Kilsyth
(population 7600, decrease 6·2 per cent.), and Denny-
and-Dunipace (population 5130, decrease 0·7 per cent.).
The population of Falkirk has approximately doubled
since 1891, of Grangemouth since 1881, of Kilsyth and
of Denny-and-Dunipace since 1851, and of Stirling since

1841. In the parishes, in 1911 14 were found to have increased in population during the decade 1901–11, and 8 to have decreased. The largest increase was found in Falkirk (5795 or 15·8 per cent.), and the largest decrease in Slamannan (1846 or 34·9 per cent.). In 1921 the parishes showing the largest increases in population were St Ninian's (964 or 7·0 per cent.), Logie (732 or 16·7 per cent.), Falkirk (340 or 0·8 per cent.), Stirling (272 or 1·3 per cent.), Airth (262 or 17·3 per cent.), and Strathblane (251 or 24·5 per cent.). The parishes showing the largest decreases were Grangemouth (763 or 3·9 per cent.), Kilsyth (688 or 6·2 per cent.), Larbert (595 or 4·6 per cent.), and Buchanan (210 or 26·4 per cent.). Each of the three county districts showed an increase in population; the Central District (population 27,475) having an increase of 4·9 per cent., the Eastern District (population 41,792) 0·1 per cent., and the Western District (11,794) 1·2 per cent. It must be understood that comparison between the present and past populations of Stirlingshire is complicated by alterations which have been made in the county boundaries. The alien element in the population is about 1 per cent.

The occupations of the people of Stirlingshire are very varied. In 1911, out of 51,551 males engaged in remunerative occupations, the largest number (11,451 or 22·2 per cent.) were engaged in the mining industry. Engineers numbered 8706, or 16·9 per cent., of whom 6420 were ironfounders. In the building trade (masons, bricklayers, carpenters, joiners, painters) were 2764, while agriculture engaged 3375. In regard to females, the census report gives no statistics about women employed in household duties for which no earnings are taken. Hence in the figures for 1911 only 13,427, or 22·7 per cent. of the females of the county, are returned as employed. Of these 3793, or 29 per cent., were domestic indoor servants. The rest were variously employed in farming, dressmaking, clerical work, teaching, laundrywork and other occupations.

10. Agriculture.

In its present state the agricultural system of Scotland dates from a comparatively recent period. During the earlier half of the eighteenth century farming was still carried on throughout the country for the most part upon the very primitive lines which had descended from the Middle Ages. Over wide areas the old "run-rig" system, with all its impediments to individual initiative and enterprise, still persisted. A great deal of the country was unenclosed, and holdings were too small to be sound economic units. Ploughs and other agricultural implements were of the clumsiest and rudest description, threshing mills did not exist, and the harvest was gathered with the sickle. Artificial fertilisers had not been invented, potatoes and turnips were unused, no rotation of crops was observed, and stock breeding was imperfectly understood. In those days cultivation did not occupy the fertile bottoms and alluvial flats, which then were wastes of dreary marsh. It clung rather to the uplands, on the bare flanks of the hills. Hence nothing is commoner throughout Scotland than to see the "rigs" of ancient cultivation high on the hill sides which are now abandoned to pasturage: a phenomenon that does not indicate a greater extent of cultivation in former days, but simply points to a time when the lower levels were swamps or peat-mosses.

Under modern conditions, agriculture has not only become a highly scientific occupation, but in Scotland it has also become highly industrialised. The farmer now conducts his business upon a purely commercial or manufacturing basis. He grows the products of his farm, plant and animal, not principally to supply his own needs, but for sale in the external market. His output of beef or milk or oats has not much more relation to his household wants than has the output of the coal-miner or the manufacturer. Such conditions make for efficiency, for the farmer has to compete in the open market just like the manufacturer or merchant. In this matter of industrialisation, the agriculture of Stirlingshire, in common with all the Lowland districts of Scotland, contrasts markedly with the agriculture of small crofters, producing for their own needs, which still prevails in remoter parts of the country.

Another fact that has had much influence upon the agriculture of Stirlingshire is the proximity of great coalfields and large industrial cities. This has caused the agricultural labourer to share very largely in the prosperity which the coal mines have brought. The farmer has to compete in the wages-market with the coal-owner and the manufacturer: not merely must he pay wages equally high, but even higher, so as to compensate for the greater dullness of country life. Only thus can he keep his labourers on the land. The result has been a great impetus to the efficiency of Stirlingshire agriculture.

Of the 288,842 acres (excluding water) in the shire, in 1922 and 1923, 113,193 acres and 113,148 acres respectively were under crops and grass, and there were in addition 151,555 and 153,332 acres of mountain and heathland used for grazing. The area under cultivation comprised 55,278 acres and 53,747 acres respectively in the two years. The following is a list of the grain crops, with the total acreage devoted to each in 1922 and in 1923: oats, 19,169 and 18,549 acres; wheat, 2194 acres in both years; barley and bere, 1024 and 998 acres. Other crops were distributed thus: turnips and swedes, 3932 and 3919 acres; potatoes, 3555 and 2865 acres; beans, 1441 and 1381 acres. Turnips and potatoes are mainly cultivated on the dryfield soil, along the valleys and on the higher ground bordering the carse. The carse itself supports most of the wheat grown in the country, and is also well adapted for growing beans. Crops of minor importance are rye and peas, while a temporary revival of flax growing, once universal in Scotland, took place during the war, owing to the interruption of the Russian flax-trade. Rye grass and other rotation grasses and clover accounted in 1922 for 21,628 acres; 159 acres were under small fruit, and 80¾ acres in orchard. In 1923 the corresponding figures were 21,371 acres, 162¾ acres, and 70¼ acres.

Turning now to livestock, the black-faced sheep is universal, and Cheviots, a less hardy stock, are also common on the lower levels. The older white-faced sheep, which required to be sheltered at night during winter, has now disappeared. In 1922 the total number of sheep carried by the county was 114,195; in 1923, 117,094.

Irish cattle, shorthorns, and crossbreeds are common everywhere, but Ayrshires, which are famous for their milk-
yielding qualities, predominate among dairy cattle on the
farm in the dryfield area. The dairy farms tend naturally
to group themselves in the south and west of the shire,
near the populated area and in touch with the railways.
The total number of cattle in Stirlingshire in 1922 was
30,435; in 1923, 31,619. In the former year the county
also contained 5176 horses, mainly Clydesdales, and 2390
pigs: for 1923 the numbers were 5070 and 3005.

The holdings in the shire in 1922 numbered 1465, and
averaged 77·3 acres in extent. In 1923 the number was
1469, and the average 77·0 acres.

11. Industries and Manufactures.

The most important industries in Stirlingshire are those
connected with its great eastern coal-field. The coal industry will be dealt with in the following chapter. Here
we consider its twin industry, the production of iron and
iron goods. The systematic development of this industry
in Stirlingshire dates from the establishment of the great
Carron Iron Works in 1759. Previous to that time little
or no cast iron was manufactured in Scotland. The idea
of founding this industry here originated with Mr William
Cadell of Cockenzie, an extensive importer of iron and
timber from Norway and Sweden. But the prime mover in
the establishment of the works at Carron was Dr John
Roebuck, a Birmingham physician, who devoted himself
to the study of industrial chemistry, and was the first to

revive the smelting of iron ore by coke made from pit coal, and also by the same means to convert cast iron into

Craigend Colliery

Carron Blast-furnaces

malleable iron. The choice of Carron as a site for the development of the new industry was dictated by its nearness to the sea, the abundance of water-power, and the proximity of iron and coal. Skilled furnacemen and moulders were brought from Birmingham and Sheffield, and John Smeaton, the famous engineer who constructed the Eddystone Lighthouse, was called in to erect blowing cylinders of a new type for the powerful blast needed to reduce iron by pit coal. Operations began with an air furnace at the close of 1759: the first blast-furnace was ready a twelvemonth later. The new undertaking throve immensely, and in 1773 received a royal charter of incorporation with a capital of £150,000. Thus a great new industry has been created in Scotland, a fact which gave an immense impetus to the general economic development then taking place.

Section of William Symington's Steamship

For many years the Carron Iron Works were the largest in the world. During the Napoleonic wars they cast the famous light guns called "carronades" after their place of manufacture. It is also interesting to note that Major Shrapnel carried out important experiments at Carron in perfecting the type of shell named after him. Since 1852 the Company has ceased, in normal times, to manufacture munitions, and is now mainly concerned with producing machinery, boilers, ranges, stoves, fire-grates, r.w. pipes, baths, cooking apparatus, electric heating and cooking appliances, and miscellaneous iron articles. In 1789 the machinery for the first practicable steam vessel was built at Carron from designs by William Symington, who also erected the first steam-blowing vessel here.

At present the Carron Iron Works occupy an area of about 55 acres, with 40 acres of water in storage dams and a similar area of open land. The main entrance is surmounted with a clock tower, having the arms of the

Aerial view of Carron Works

Company—two crossed cannon, with motto *esto perpetua*. In normal times the Company employs between 5000 and 6000 hands. In 1845 about 8000 tons of pig iron were smelted; in 1878 the figure had increased to 41,343 tons. The Company owns extensive lands in the neighbourhood, as well as properties in Airth, Grangemouth, Bothkenner, Denny, Muiravonside, Kilsyth, and Slamannan, and also in Fife and Cumberland. These lands are held chiefly in connexion with the Company's mineral undertakings. The Company maintains a fleet of cargo-steamers sailing between Grangemouth and London, connecting with Glasgow and the west of Scotland and Belfast and Northern Ireland.

There are extensive iron foundries at Falkirk, which is the chief centre of the light casting trade in Scotland, and at Denny, Grangemouth, Kilsyth, and Stirling. The light casting trade now totals 30 foundries, with about 7500 employees and an annual output of 50,000 tons. The manufacture of stoves, grates, and kitchen ranges is extensively carried on at Bonnybridge. In St Ninian's and district the nail-making industry has long been established. The scattered forges where hand-made nails were produced formed a picturesque feature of these villages, but are now seldom to be seen, as the industry is concentrated in factories equipped with modern machinery. Chemical works are found at Denny, Falkirk (where explosives are manufactured), and Lennoxtown. The dyeing industry is established at Bridge of Allan. At Lennoxtown, Denny, and elsewhere are printworks and bleachfields. The woollen industry is an important one in Stirlingshire, and is an instructive example of how an industry may be perpetuated in a locality after the original conditions that led to its establishment have vanished. The wool trade in Stirlingshire originated through two main causes: the proximity of great sheep pastures in the Ochils and Campsies, and the abundance of water-supply in the burns that hurry down those hills. Once the industry had taken root, the development of the coal-fields supplied a more effective motive-power and ensured its permanence. Its chief centres are now at Cambusnethan, Stirling and Bannockburn. The cotton industry at Balfron is not now of great importance. There are numerous distilleries and breweries in the shire, notably at Falkirk, Bonnymuir, Camelon and Gargunnock. Originally Kippen was a centre of the whisky trade. The timber industry is carried on at South Alloa and Grangemouth. At Bonnybridge, Falkirk, Grangemouth, Rough Castle and Castlecary there are brickworks. The yellow clay of the carse makes excellent bricks and tiles, but is not now in use for this purpose. Tile-drains for agriculture are still made. Denny is one of the chief centres of the paper-making industry in Scotland. There are four extensive mills, the largest employing 500 workers. There is also an important paper-works at Bridge of Allan.

The connexion of Stirlingshire with shipbuilding is an ancient one, James IV having had his naval arsenal and docks at Airth. This industry is now located at Grangemouth, where the first steam vessel was launched in 1839. The huge docks here are now the property of the London, Midland and Scottish Railway. These docks cover about 60 acres, with 50 acres of timber basins and 50 acres of approaches and connecting channels. The series of docks, basins, and channels stretches a mile and a half out into the Forth. Upwards of 1000 men are employed more or less steadily in connexion with these docks.

The fisheries of Stirlingshire are unimportant. On the Forth Salmon fishing is still continued by the primitive method of coble and net. Twice a year great numbers of sparling (p. 38) swarm up the river and are caught and sold in the local market. In the days of the poet Dunbar the "sperling" of the Forth were famous. The pearl mussel fishing of the Forth and Firth still affords occasional employment for a number of men.

12. Mines and Minerals.

The coal mines of Stirlingshire, with their associated industries, form the most important economic unit in the county. The Stirlingshire coal-fields and their accompanying beds of oil-shale and iron-ore form part of the great central mineral field of Scotland, which extends in a broad band right across the country, between lines drawn roughly from Ardrossan to Elie and from Dalmellington to Tranent. Within this area five separate coal-fields may be distinguished, namely (1) the Ayrshire field; (2) the Lanarkshire field, which yields nearly half the total Scottish output; (3) the Stirlingshire field, including part of Clackmannan; (4) the Lothian field; and (5) the Fife field. Economically no less than geologically the Stirlingshire coal-field must be regarded as a north-eastern extension of the Lanarkshire basin. It must always be remembered that in the mining industry of Great Britain the Scottish coal-fields bear only a relatively small part, about 13 per cent. in value of the total production. During the period 1907–11 the average annual value of the Scottish coal production was about £15,000,000, as against £97,000,000 for England and Wales. In tonnage the Scottish output was about 15 per cent., and of this Stirlingshire's share was about 6 per cent. So also in the associated iron industry the Scottish share is comparatively unimportant, about 4 per cent. of the total British production. More than one-third of Scottish iron-ore is extracted from mines that also produce coal. In iron production Stirlingshire ranks eighth in order of output among the Scottish counties, those out-distancing it being, in order of merit, Ayr, Lanark, Renfrew, West Lothian, Dumbarton, Fife, and Midlothian.

Scottish coal is of the soft or bituminous variety : the hard, smokeless coal known as anthracite, produced in the South Welsh coal-fields, is scarcely known in Scotland. It was worked, however, in the now exhausted Slamannan coal-field, where the coal in the Longriggend district had been burned by an igneous intrusion : but this Slamannan anthracite was of little commercial utility. Generally speaking, Scottish coal is less valuable than English, but it enjoys a special advantage in its geographical position, the coal being found in a localised area with easy communications, and in close touch with the fertile Lowlands. The influence which the coal industry has thus exercised on other Scottish industries has already been discussed. It must always be remembered that while Scotland is still a predominantly agricultural country—the value of her farm produce greatly exceeding that of her minerals—yet the coal mines have exercised a wholly disproportionate influence on her modern development, raising the country from poverty to comparative wealth, and effecting a radical change in the distribution of its population. A notable feature in the growth of the mining industry is the effect it has exercised in stimulating the importation of raw material from foreign sources. Thus the presence of iron in the Scottish coal-fields led to the foundation of the iron industry, whose present huge dimensions have totally out-run the local supplies, so that vast quantities of iron are now imported.

One interesting feature about the towns in Scotland which owe their prosperity to coal and its associated industries is that they are for the most part ancient centres to which mining has brought great expansion in modern times. A striking example of this is Falkirk, a burgh of high antiquity and formerly a great agricultural centre, but now depending entirely on coal and the manufactures that coal has made possible. Stirling itself shows the same phenomenon in a lesser degree. In England, on the other hand, most of the big industrial towns have arisen on new or formerly unimportant sites. This contrast is of course due to the fact that the Scottish coal-fields occur in that part of the country which from other causes has always been the most densely populated.

The intimate local connexion that exists between the mining and other industries in Stirlingshire is well illus-trated by the now extinct nail-making industry of the St Ninian's district (p. 58), which owed its prosperity entirely to the neighbourhood of a good smelting coal worked from the Bannockburn Main or "Bottom Coal" seam.

The Stirlingshire coal-fields show very different stages in the development of the industry. Thus the Slamannan area, long famous for its excellent steam coals, must now be regarded as exhausted. Its four principal seams are worked out, and the thinner underlying seams offer no economic possibilities. Hence the district is becoming depopulated (p. 50). On the other hand the Falkirk area, despite the complete exhaustion of its most profitable seam and the great depletion of the two others next in value, still presents important reserves in its numerous thinner seams, which being less profitable have hitherto escaped working in their deeper levels. The progressive exhaustion of Scottish coal and the development of coal-cutting machinery is likely to bring these seams into future prominence.

13. History.

The activities of the Romans in Stirlingshire will be dealt with under the Roman Wall (p. 80). In the fifth century Stirling seems to have been part of the Britonic kingdom of Strathclyde. Christianity was introduced about 400 by St Ninian, the apostle of Strathclyde. He founded the church called after him at St Ninian's, which originally was known as *An Eaglais*, or Eccles, "The Church," indicating that there was then no other Christian site in the district. Places named after this saint existed also at Stirling, where his well still flows copiously, and at Campsie. In the seventh century the work of St Ninian was completed by St Blane of Bute and St Modan of Rosneath. St Kentigerna, a sister of St Kentigern, founded a nunnery on Inchcailloch ("Island of Nuns") in Loch Lomond, where she died in 734. Towards the end of the seventh century the English kingdom of Northumbria appears to have pushed its frontier north and west as far as Stirling; but, after the disastrous defeat of King Egfrid by the Picts at Dunnichen (May 20th, 685), the Anglian outposts shrank back to the Pentlands. The final Teutonisation of the district took place under Anglo-Norman auspices in and after the twelfth century.

Stirling Castle, around which the history of the county centres, emerges as a royal residence under Alexander I, who died here (1124). So early as 1119 the town was a royal burgh, one of the Court of Four Burghs which later grew into the present Convention of Royal Burghs. Under the humiliating Treaty of Falaise (1174), its castle received an English garrison. William the Lion was much attached to the castle, made a royal park beneath it, and died within its walls (1214). Both Alexander II and Alexander III spent much time at Stirling, and by the close of the thirteenth century it was unquestionably the leading royal residence.

During the twelfth and thirteenth centuries the Celtic Church was reorganised by Norman ecclesiastics imported by the Scottish Crown. Stirling now formed part of the diocese of Glasgow, founded by David I when Prince of Cumbria, about 1116. Besides the secular clergy, establishments of regular clerics were introduced, such as the Augustinian Abbey of Cambuskenneth, often known as the Abbey of Stirling. It was founded by David I in 1147, and colonised by monks from Aroise in France. In the town of Stirling were establishments of Dominicans (Blackfriars) and Franciscans (Greyfriars). At Manuel, on the eastern border of the county, a priory of Cistercian Nuns was founded by Malcolm IV in 1156.

The Wars of Independence broke heavily on Stirling-shire. In 1291 the castle was yielded to Edward I, but was returned next year to Balliol. In 1296 it again received an English garrison. At Stirling Bridge (Sept. 11th, 1297) Wallace gained his great victory over Cressingham, as a result of which the fortress passed back into national keeping. Next year at Falkirk (July 22nd, 1298) Edward in person reversed the scale of fortune. Thereafter the castle was dismantled by Wallace, but being rebuilt by Edward was captured by the Scots in 1299. In 1304 it sustained a famous siege at the hands of Edward, and was heroically defended by Sir William Oliphant. So intent was the English king on its capture that he ordered the churches of Perth, St Andrews, Brechin and Dunblane to be stripped of their lead to provide weights for his siege engines. At last (July 24th, 1304) the garrison, less than 150 strong, surrendered. Edward's queen, Margaret of France, watched these operations from an oriel specially constructed in her house in the town of Stirling.

In 1313 the castle was blockaded by Edward Bruce, who made the celebrated bargain with the governor that it should be yielded if not relieved by June 24th of next year. That compact resulted in the epoch-making battle of Bannockburn (June 23rd–24th, 1314), in which the great English army under Edward II was defeated by the Scots under Robert I. Stirling Castle surrendered next day; and, after Bruce's usual policy, its walls were thrown down.

At Cambuskenneth Abbey in 1326 King Robert convened a notable parliament, the first to which burgh representatives were summoned. During the minority of

David II, Edward III renewed aggression upon Scotland. In 1336 Stirling Castle again admitted an English garrison, when the walls destroyed by Bruce were rebuilt. In 1341–2 it was besieged by the Scots, and forced by starvation to capitulate. During this siege cannon were employed for the first time against the castle. This was the last occasion when Stirling Castle was in English keeping, though an unsuccessful attempt on it was made by Richard II in 1385.

In May, 1425, Murdoch, Duke of Albany, formerly Regent of Scotland, was executed, with his two sons, on the Heading Hill of Stirling, by command of James I. On February 22nd, 1452, Earl Douglas was stabbed in the castle by his host, James II. In revenge the Douglases burned the town of Stirling. James III was born in the castle, and outside its walls he sustained his fatal defeat from the rebel barons at Sauchieburn (June 11th, 1488). This battle is memorable as the only engagement in which Highlanders and Borderers faced each other in arms. The slain king was buried in Cambuskenneth Abbey, beside his queen, Margaret of Denmark (p. 91).

James IV and James V both resided mainly at Stirling, and the present aspect of the castle is largely due to their buildings (pp. 96–7). The ancient connexion of Stirling with St Ninian was recalled on the foundation of the Chapel Royal within the castle in 1503, when Pope Julius II appointed as its Dean the Bishop of Galloway, whose cathedral seat was at St Ninian's ancient settlement of Whithorn. It was from Stirling Castle that James V set out on his wanderings incognito among his humbler subjects, and from the hollow behind the castle he took his popular name, "Gudeman o' Ballengeich." Here also his infant daughter, Queen Mary, was crowned (Sept 9th, 1543), and the walls of the castle mainly sheltered her for the first six years of her troubled life, until she was sent away to France. On the day before the little Queen was crowned, Arran "the sillie Regent," dominated by the

Gothic Offertory Coffer from Cambuskenneth Abbey

forceful Cardinal Beaton, did humble penance in Stirling Church for his apostasy to the Reform movement.

James VI was baptised in the castle on December 17th, 1566, and in the parish church of Stirling he was crowned on July 29th next year. The first twelve years of his life were spent within the castle, where his stern tutor was the famous Latin scholar, George Buchanan (pp. 120–1). On

5-2

April 7th, 1571, Archbishop Hamilton, last Roman primate of Scotland, was hanged at Stirling market cross. He was tried and executed on the same day, and was the first bishop in Scottish history to die at the hands of the law. On Sept. 4th, 1571, a skirmish was fought in the streets of Stirling between the partisans of the deposed Queen Mary and those of her son's government, and the Regent Lennox was slain. On March 17th, 1578, the town saw a bitter fight between the followers of Lord Crawford and those of the Lord Chancellor Glamis, who was killed in the fray. On April 17th, 1584, Stirling Castle was seized by the Ruthven Raiders, but was speedily recovered by the king. Next year on Nov. 4th they again captured it, and dictated their terms to the unlucky James. The king's eldest son, Prince Henry, was born at the castle (February 19th, 1594), and on August 30th he was baptised in the Chapel Royal, rebuilt for the occasion. He was the last Scottish prince to be brought up in the castle. It was visited again by James in 1617, and by Charles I in 1633. In 1650 Charles II, on his way to defeat at Worcester, passed a few nights in the castle, and with this event its history as a royal residence closes.

Stirling Castle played no great part in the Civil War, but at Kilsyth the Royalist leader Montrose gained his crowning victory (August 15th, 1645), which briefly threw all Scotland at his feet. In July 1651 Cromwell and David Leslie faced each other in the Torwood near Stirling, and Callendar House was burned. On August 14th the castle yielded to Monk after a short cannonade. At a conventicle held in the Torwood, in Sept. 1680, the Covenanters under

Queen Mary and her son

Donald Cargill excommunicated the king, his brother James, Duke of York, and the chiefs of the Government in Scotland. During the Jacobite revolt of 1715, Stirling Castle and Bridge were strongly held for the Crown. In the second rebellion (1745) the town was easily captured by Prince Charles, but the castle withstood an ill-planned siege. On this occasion the south arch of Stirling Bridge was blown up to obstruct the rebels. Hawley's attempt to raise the siege resulted in the Jacobite triumph at Falkirk (Jan. 17th, 1746), which is the last incident in the military history of Stirlingshire.

Owing to the vital importance of Stirling Castle and Bridge (p. 2), the surrounding district is *par excellence* the battlefield of Scotland. Seven important battles, six in the shire, have taken place round Stirling since the Wars of Independence: Stirling Bridge (1297), Falkirk (1298), Bannockburn (1314), Sauchieburn (1488), Kilsyth (1645), Sheriffmuir (Perthshire, 1715), and Falkirk (1746).

In 1773, owing to municipal corruption, the town of Stirling was deprived of its privileges, but these were restored in 1781. An event in its modern history which aroused intense interest was the execution (Sept. 8th, 1820) of Andrew Baird and John Hardie, leaders in the "Radical Rising," who were captured at the "Battle" of Bonnymuir (April 25th, 1820).

14. Antiquities—(i) Prehistoric.

The rich plain of Stirlingshire has been inhabited from remote times. In the clay subsoil of the carse-lands near Stirling remains of whales have been discovered, which

must have stranded there when the clay was a slimy mud
at the bottom of a shallow inland sea. Vast topographic
changes have since occurred. Upheaval caused the sea to
retire, and the exposed carse-land became clothed with
stately forests. Of these the fallen trunks, and roots still
in situ, are often discovered. The forests in their turn gave
place to wide mosses, which persisted until cultivation
drained them within recent times. In Blair Drummond
Moss, on the Perthshire side of the Forth, a "corduroy"
road, formed of logs from this ancient forest, was discovered.
Many logs showed marks of the axe, and the causeway
has been attributed to Roman legionaries. If so, we have
a lower date for the period of forest growth. But clearly
the epoch when whales were stranded on the mudflats of
an inland sea, where afterwards these forests throve, was
infinitely more remote. Yet along with the cetacean skele-
tons, implements of deerhorn and wood were found, which
must have been used by primeval hunters to strip the bones
of their blubber (p. 46). And on several occasions "dug-
out" canoes have been found deeply embedded in the
marine clay of the Carse of Falkirk. These may have been
the vessels used by the primitive hunters who spoiled the
carcases of the whales stranded on the shores of that ancient
sea. "Kitchen-middens," heaps of oyster and mussel shells,
found at various places along the 50 foot beach, indicate
another way in which these earliest human inhabitants of
Stirlingshire eked out a precarious existence.

Antiquities of the Stone and Bronze Ages are common
in most parts of the shire. Stone axes have been found at
Kippen, Alva, Polmont, Camelon, Stirling, and Abbey

Craig. A stone hammer, perforated for a haft, was found at Park of Keir, Bridge of Allan. The county has also yielded some fine examples of the beautiful, leaf-shaped swords of the Bronze Age. Specimens have been unearthed near Carron, at Ballagan, and in Poldar Moss. A fine bronze spear-head was found in a moss near Stirling, and another was obtained near Falkirk. A flat bronze axe was

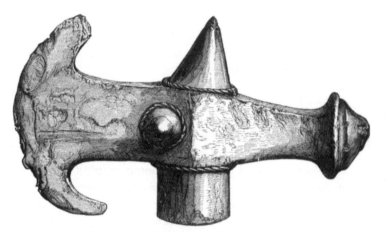

Bronze Axe-head found at Bannockburn

found at Airth, and one of the flanged variety near Stirling Bridge. Perhaps the finest bronze axe in Scotland was dug up on the field of Bannockburn. It is probably of the Early Iron Age.

No graves undoubtedly belonging to the Stone Period seem to have been recorded from Stirlingshire. Interments of the Bronze Age are very common. At the beginning of this period the dead were inhumed in a crouched position in short stone cists, but later cremation set in, the ashes being gathered into a cinerary urn. In cist burials urns of special types are usually found, which are believed to have contained food for the journey to the after-world. They may be either "food vessels" or "beakers," the latter being taller and narrower than the former. Two cairns in Baldernoch parish contained cists with human bones, and in one cist also urns. One of the largest cairns in Stirlingshire, the "Ghost's knowe" at Craigengelt, was demolished in 1839. It was circular, 300 feet in circumference and 50 feet high. Around its base were placed, at regular intervals, 12 large stones. Within was found a chamber, containing a skeleton, alongside which lay a golden horn or cup, a polished stone hammer, a stone knife and a gold finger ring. An admirable example of a short cist burial was found at Coneypark in Cambusbarron. The cist measured 3 feet 7 inches by 2 feet 10 inches, and contained a skeleton with the very round head-form usual in such burials (p. 47). From other cists found at Cambusbarron urns of both types, food vessel and beaker, were recovered.

Isolated cinerary urns have been reported from numerous localities. Near Cambusbarron a Bronze Age cemetery was explored. It yielded four cinerary urns, a perforated stone hammer and a fragment of bronze. Quite recently two graves were found near Camelon, one of which contained incinerated remains, while the other yielded a food vessel and unburnt bones. At Dungoyach, near Duntreath Castle, remains of a stone circle exist. Such constructions are usually found to contain burial deposits of the Bronze Age. Two standing stones may be seen in the park at Airthrey Castle, and other remains of stone circles exist in the county.

Bronze Age Sepulchral Pottery from Camelon

A thick penannular gold ring found at Bonnyside is probably of the Bronze Age. Relics of the prehistoric Iron Age are rare, owing to the corrosive nature of the characteristic metal. Harp-shaped brooches of Late Celtic design have been found near Castlecary and Falkirk. An outstanding example was dug up on the estate of Polmaise. It is of brass, with a copper pin, and has the usual trumpet-shaped and crescentic ornament. A splendid silver brooch, made of thin sheets of hammered bronze with zoomorphic and interlaced patterns in gold, and set with amber, was found at Dunipace. Fragments of pottery of the Iron Age

Brooch found at Dunipace

were dug up in 1915 during the making of entrenchments at Bantaskine. A grave at Sauchie contained a quantity of silver bracteates, or Roman coins strung on a necklace.

On several hill tops the banks and ditches and drystone walls of early fortifications may be traced. One of these hill forts girdles the summit of Abbey Craig, enclosing the Wallace Monument. Another, with three concentric walls, crowns Blairlogie Hill. A number of earthworks

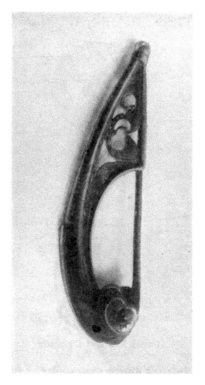

The Polmaise Brooch

are found upon the flanks of the Campsie Fells. Meikle Reive and Maiden Castle, upon the south face of Lairs Hill, are good examples.

The brochs, a class of fortified structure peculiar to Scotland, are circular towers of dry-built masonry, en-

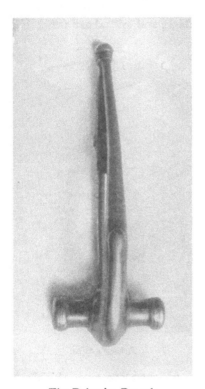

The Polmaise Brooch

closing an open court. Their thick walls are hollowed into galleries, reached by a stair which winds round the tower to the parapet. The relics found in such structures prove them to belong to the Iron Age. Remains of a broch, known as the Tappock, stand on the summit of the Torwood. Its court is 35 feet in diameter, within a wall 15 feet thick, and still about 6 feet high. A narrow

Arthur's O'on

door gives access to the court, from which a passage leads to the winding stair. The relics found on excavation included pottery, an iron hammer and axe-head, whetstones, whorls, lamps, querns, personal ornaments of stone, and charcoal. Three large blocks of sandstone bore the cupmarks which are a puzzle to archaeologists. Examples of these mysterious sculpturings may be observed on the bare heath-rock of Craigmaddie Moor, near the "Auld Wives' Lifts" (p. 30).

Perhaps the most widely discussed of Stirlingshire antiquities has been the famous "Arthur's O'on," which stood near the Carron Iron Works. It was destroyed in 1743 to provide stones for a mill-dam. It was a circular

The Stirling Stoup

beehive structure, about 90 feet in basal circumference and 22 feet high, with a round opening 12 feet in diameter at top. There was an arched doorway 9 feet high, and above it a window with sloping sides. The structure was built of hewn and coursed freestone, but without mortar. Older antiquaries regarded this as a Roman building, but the details suggest that it was Celtic.

In the Museum of the Smith Institute at Stirling is a well-arranged collection of prehistoric antiquities, many of which are local. The Museum also contains an extensive series of neo-archaic domestic and burghal antiquities, including the famous "Stirling Stoup," dating from 1405–21, which is the oldest Scottish measure extant, and formed the standard from which all Scottish measures were struck.

15. Antiquities — (ii)   Roman — the Roman Wall.

The chief antiquity in Stirlingshire is the Roman Wall, built by Lollius Urbicus, governor of Britain, about A.D. 143,

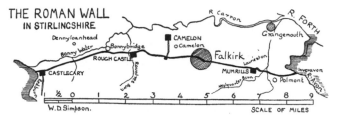

in the reign of Antoninus Pius. It represents an expansive movement from the frontier drawn by Hadrian in his wall between the Solway and the Tyne, built about A.D. 122. The advanced line was held until A.D. 181, when a great Caledonian inrush swept the Romans out of Scotland: and the Emperor Severus, after his punitive campaign in Caledonia in A.D. 208–10, fixed the frontier finally at Hadrian's barrier. The Antonine Wall occupies the line of

an earlier chain of posts, established by Julius Agricola in
A.D. 80. It stretches from Bridgeness, near Carriden on
the Forth, to Old Kilpatrick on the Clyde, and is about
36½ miles in length. It enters Stirlingshire near Inveravon
and leaves it just west of Castlecary. Nearly 11 miles of
the Wall are thus in our county.

The fortified line occupies a position of great strength,
in its eastern section overlooking the Carse-lands bordering
the Forth. It consists of a wall 14 feet wide, carefully
built of coursed sods resting on a stone bed, and was
probably 10 or 12 feet high. Outside the wall is a dry
ditch, sometimes as much as 40 feet wide and 12 feet
deep. Between ditch and wall is a berm, and beyond the
ditch a *glacis* is formed of the upcast. At intervals along
the wall, sod-built ramps abut on its inner side, doubtless
signal stations whence bale-fires flashed afar the news of
approaching war. East of Falkirk the wall consists not of
sods but simply of earth. Everywhere the defensive line
chooses the most advantageous ground, even where this
has involved a rock-hewn ditch.

The garrison was stationed in forts, probably 19 in
number, placed on an average about two miles apart. Of
these Mumrills, Rough Castle, and Castlecary, as well as
two whose precise site is unknown, are in Stirlingshire,
which also contains the post of Camelon, north of the
Wall. A military way connected the stations. The troops
were normally light-armed auxiliaries, the heavy legions
being in the fortresses of Chester, York, and Caerleon,
further south. The legions were a field force employed
for a serious campaign, but too valuable for the rough and

tumble of frontier defence. But they were utilised in building the Wall, inscriptions revealing that all three British Legions—2nd, 6th and 20th—shared in this work. Inscriptions also show that along the Wall were stationed corps recruited from Holland, the Rhine valley, Gaul, Spain, Thrace, and even distant Syria. They were mainly infantry, but a cavalry regiment held Mumrills, overlooking the Carse of Falkirk, a wide plain suitable for horse.

As an example of a Roman settlement in Stirlingshire, we may take Camelon, which has been excavated. It is situated some 1100 yards north of the Wall, with which it is connected by the military road that led by Stirling to the great camp at Ardoch in Perthshire. Camelon is strongly posted on the edge of steep banks overhanging the Carron. It comprised a main camp of nearly six acres, and an annexe of more than eight acres to the south. Both were girt by strong entrenchments. In the north camp, from an entrance on each face, streets of hard gravel, paved near the gates, bisected the enclosure. At the centre were the "Principia" or Headquarters. The remainder of the camp was laid out in streets of barracks, storehouses, and officers' quarters. All the houses were built of hewn stone. Some living rooms had floors of concrete or flagstones, and were warmed by "hypocausts"—the flooring being supported on dwarf pillars, forming a space from which heated air was passed up through the walls. Fragments of window glass and red roofing tiles have been found, while moulded stones and the base of a small statue give evidence of some architectural pretension. An admirable drainage ensured the health of the garrison. The annexe

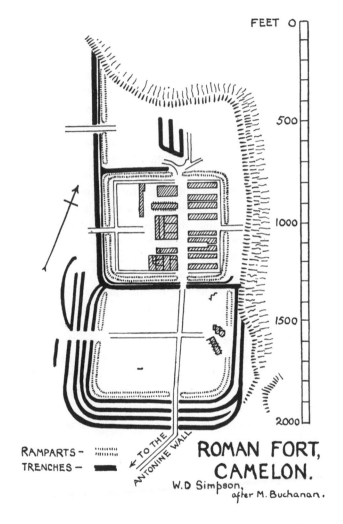

FEET 0

500

1000

1500

2000

RAMPARTS –
TRENCHES –

TO THE
ANTONINE WALL

ROMAN FORT,
CAMELON.
W.D Simpson,
after M. Buchanan.

6-2

doubtless sheltered the civil populace attracted by the soldiery. Here was a fine suite of baths.

From the articles found some idea may be obtained of the life led by the garrison. Among them are many fragments of pottery, including the red, lustrous ware usually called "Samian," adorned with reliefs sometimes depicting gladiatorial combats and scenes from the chase; bronze and clay lamps; beads and bangles in stone and glass; studs and brooches in bronze, sometimes richly enamelled or silvered; harness-mountings; and various tools and weapons—pickaxes, adzes, hammers, weaving-combs, and spear-heads. A number of pieces for a game like draughts show how the soldiers whiled away their leisure. An altar was found but without dedication. One of the most interesting discoveries is a stone with the name of the 20th Legion, a detachment of which must have built the fort. Coins from Vespasian (A.D. 69–79) to Marcus Aurelius (A.D. 161–180) give us insight into the periods of occupation of a settlement which has been described as "a prominent post, a long maintained centre of Roman activity." This fort is of special interest because it was here that the spade first revealed the presence at a specific site of Agricola's army. Much of the Samian ware is distinctively first-century in type, and more than half the coins belong to the Flavian dynasty.

The stations at Rough Castle and Castlecary, upon the Wall, have also been excavated. The former occupies a site of great strength, and was defended by powerful earthworks. A remarkable feature is the defensive pits or "lilia" outside the Wall. These were probably the handiwork of

Stone, found at Mumrills, commemorating a soldier
of British birth

Lamp found at Castlecary

Agricola, and had been partly buried by the upcast of the Antonine ditch. Within the fort were the usual head-quarters building and granary. At the south gate the pavement was deeply worn into ruts by wheeled traffic. Outside the fort a large annexe contained baths. The relics were

Castlecary, inner face of south wall of "Principia"

scanty, but included the fragment of a life-size imperial statue, and a stone recording the building of the Principia by the 6th Cohort of Nervians.

Castlecary, like the station of Balmuildy north of Glasgow, had a stone wall with angle towers. Here also was an annexe. The interior buildings were of the usual kind, including the Principia, a granary, latrines, and baths, the furnace of which was still charged with soot. Among the relics were a statuette of Fortune and a beautiful little intaglio in carnelian, representing Jupiter and his eagle. A number of sandals were found, some from the feet of women and children. Inscriptions revealed the presence of the 2nd, 6th, and 20th Legions and the 1st Cohort of Tungrians. This station also stood on the site of an Agricolan fort, and was the strategic centre of the Wall. From Castlecary a military way appears to have run south to join, near Carluke, the great trunk road from Carlisle, by Netherby and Birrens, to the western end of the frontier.

The fort at Mumrills is peculiar in that its rampart, like the Wall itself here, is made of earth, not sods. A remarkable kiln for burning bricks was found close to this fort. Excavations now in progress here have revealed "a headquarters building probably as large as, if not larger than, any other found in Britain."

In addition to their military works the Romans have left traces of their influence upon Stirlingshire in the numerous coins which, singly or in hoards, have been turned up at various places. Apart from those discovered at Roman sites, these coins have been found at Torrance, Campsie Glen, Drymen, and St Ninian's. The Drymen find, made in an old quarry near the Endrick Water, was a small hoard, comprising two gold coins of Nero and Trajan. It must not be imagined that coins found on non-Roman sites were dropped or concealed by Roman hands. Archaeological evidence suggests that Roman coins were regularly used as currency by the natives of Scotland during and after the imperial occupation.

16. Architecture—(i) Ecclesiastical.

The buildings of the early Celtic church were dry-built stone cells with beehive roofs. None survives in Stirlingshire. In the twelfth century the ecclesiastical organisation of Scotland was refashioned upon Anglo-Norman lines, and along with the new church system the characteristic Norman or Romanesque architecture was introduced. Norman churches are known by circular piers with cushion or scalloped capitals and square abaci, by round arches with chevron, billet, and other mouldings, and by broad, flat buttresses. The style is characterised by massive strength. In its later development foliage appears on the capitals. Transitional work of this kind exists at Airth Church. Fragments of a church on Inchcailloch, Loch Lomond, also exhibit late Norman features.

At the end of the twelfth century Norman architecture gave place to the oldest true Gothic, known as the First Pointed style. The buildings are loftier, with slender columns often clustered, and pointed arches richly moulded with bold rolls between deeply undercut hollows. Tracery emerges through the coupling of lancet windows under a general arch, the tympanum being pierced with circles, trefoils, and quatrefoils. Buttresses are narrower and bolder. The typical enrichment is the dog-tooth. In Scotland the round arch is often retained throughout the pointed styles. Stirlingshire has a fine specimen of First Pointed architecture in Cambuskenneth Abbey.

After the thirteenth century Scotland, owing to the breach with England, sought models from her ally France.

Scottish churches of the later Middle Ages show many French features, particularly in the flowing lines of their tracery, so different from the rigid Perpendicular style then prevalent in England. But the later Scottish Gothic is no mere copy of foreign patterns. Its foreign features are not mechanically imitated but organically assimilated. It displays a remarkable tendency to revert to the round arch and plain cylinders of Norman work. The militarisation of Scottish life, owing to the chronic war with England, had a curious result in giving these later churches a half-baronial appearance. They are furnished with the battlements, crow-stepped gables, heavy vaults, and saddle-backed stone roofs of the feudal castle. All these features are instanced in Stirling Church (pp. 91–2).

The late Norman fragments in the ruined church of Airth consist of two cylindrical piers with foliaged capitals and square abaci supporting a plain arch, slightly pointed. These early remains are built into the burial aisle of the Bruces of Dunscrub, which bears the date 1614. The Elphinstone Aisle was erected in 1593, as appears from a dated coat of arms. The Airth Aisle was built towards the end of the fifteenth century. It contains the effigy of a lady, wearing a coif and draped with a coverlet. At her feet crouch two quaintly-wrought hounds. A picturesque, tall square belfry tower is inscribed IVLY THE 15, 1647. In the grave-yard is an iron mortsafe, dated 1832.

Cambuskenneth Abbey is now reduced to mere foundations, except three fragments: a detached bell-tower at the north-west corner of the church; the west doorway; and a ruined portion of the conventual buildings, overlooking the Forth. The cruciform church measured 178 feet in length within the walls. South of the nave was the cloister-garth, beyond which lay the refectory. On the east side of the garth was the chapter-house. Though much decayed,

Cambuskenneth Abbey

the west door of the church is a good example of vigorous thirteenth century work. But the glory of the Abbey is its square bell-tower, about 70 feet high to the modern parapet. The basement has a groined vault, with a circular "eye," or aperture for hoisting the bells, and slots for the ropes. In front of the High Altar of the church lie James III and his wife, Margaret of Denmark, beneath a tomb erected by Queen Victoria in 1865.

The Parish Church of Stirling comprises west tower, nave, transepts, and choir. Nave and choir have aisles flush with the transepts, so that the cruciform plan is not revealed externally. Over all the church measures about 200 feet by 55 feet within the walls. Choir and nave now form separate east and west churches. The nave was built between 1414 and 1456. It has a clearstory of round arched windows on the south side only, and five bays of piers, cylindrical except the eastern pair which are clustered. The arches are pointed and plainly moulded. The roof of native oak has been described as "the finest open-timber roof of the Middle Ages possessed by any church in Scotland." In the aisles are groined vaults, traceried windows, and bold buttresses. Four great piers at the crossing, the eastern pair richly clustered, indicate an intended central tower. At the north-east corner of the nave is the groined chapel of St Andrew, erected about the end of the fifteenth century. In its lower part the west tower is contemporary with the nave, but was heightened when the choir was added. This lower stage is groined, with a circular "eye." The tower has an embattled parapet, and similar defences run along the nave. The upper walls of the tower are pitted by shot from the castle during the sieges of 1651 and 1746.

To the nave a choir was added in 1507–55. It consists of three bays of clustered piers supporting richly moulded pointed arches. The groined aisles have large windows with flowing tracery and buttresses with canopied niches. The clearstory windows are of two lights, round arched. The roof is higher than that of the nave, and lies between crow-stepped gables. From the east gable projects an apse of five sides with buttresses and battlements. It has a pointed vault with ribs, carrying a steep stone-slabbed roof. The east window has tracery of a somewhat Perpendicular design.

The Church of Stirling

This apse is a most effective composition. The "master-mason" was John Coutts, who received the freedom of Stirling on October 21st, 1529. His work at this, "the last of the great medieval churches erected in Scotland," shows him to have been "a man of original genius, one of the greatest of the master-masons of the later Middle Ages."

A few fragments of ornate fifteenth century work remain in the ruined church of St Ninian's. The tall, quaint west tower dates from 1736. This church was used as a magazine by the Jacobites in the "Forty-five," and was wrecked by an explosion.

17. Architecture—(ii) Castellated.

The castle, or private stronghold of a feudal lord, erected to dominate the fief of which it formed the administrative centre, is totally different from the earlier brochs and hill forts, which were public refuges of the primitive tribes. Castles were introduced by the Anglo-Norman barons who in the twelfth century settled in Scotland and feudalised her institutions. The Norman castle was an earthen mound, enclosed by a ditch and having usually a "bailey" or entrenched courtyard. The top of the mound was girt by a palisade and contained the lord's residence. The bailey, similarly strengthened, sheltered the hall, chapel, and lesser buildings. Such castles were known as "mottes." The King's Knot, south of Stirling Castle, was a structure of this type. It was much altered in 1490–1503, when James IV, in connexion with a "pleasaunce" which he laid out here, converted the motte into a "Table Round." East of Seabegs House is a circular mound, enclosed by a ditch. A charter of 1542 specifies "lie mot de Seybeggis," proving that this mound is the remnant of a Norman castle. Another example is the motte of Balcastle, west of Kilsyth. But the most remarkable specimen of these early fortresses is Sir John de Graeme's or Dundaff Castle, near Fintry. The motte is square, measuring 74 feet either way, and is girt by a deep ditch about 30 feet wide, having a high counterscarp. Attached to the motte is a bailey with ruins of stone buildings.

The historical importance of Stirling Castle, the strategic centre of Scotland, has already been discussed (p. 2); but no less interest attaches to the fortress architecturally. The

Stirling Castle from the south-west

immense rock, rising 250 feet above the plain, carries a mass of buildings unique both in their military and domestic features. From the town the approach lies across a spacious esplanade, on which is the national memorial to Bruce, a mailed statue looking towards Bannockburn. At the far end of the esplanade a broad, deep fosse isolates the castle. Beyond are batteries commenced about 1690 and enlarged before the "Forty-five." Recent excavations have disclosed

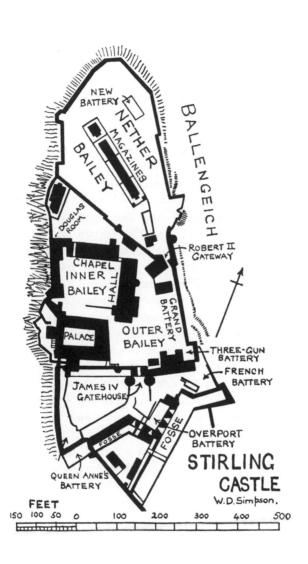

NEW
BATTERY

BALLENGEICH

NETHER
BAILEY

MAGAZINES

DOUGLAS
ROOM

ROBERT II
GATEWAY

CHAPEL
INNER
BAILEY

HALL

GRAND
BATTERY

PALACE

OUTER
BAILEY

THREE-GUN
BATTERY

FRENCH
BATTERY

JAMES IV
GATEHOUSE

OVERPORT
BATTERY

FOSSE

FOSSE

STIRLING
CASTLE
W. D. Simpson.

QUEEN ANNE'S
BATTERY

FEET

150 100 50   0   100   200   300   400   500

a series of older chambers underneath. At the east end of these is the Spur or French Battery, erected by Mary of Lorraine in 1559. A bridge spans the fosse, and two gateways, the upper inscribed "A. R." (Anna Regina), lead through the batteries. Behind is the medieval curtain wall and entrance. The gatehouse has a central arch and lateral posterns, and is defended by two round towers. Originally there were four, and the base of one of the missing pair can be seen on the left. The Treasurer's accounts for the "Gate-tower" and "Fore-entry," with a "great portcullis," show that these works were erected by James IV, 1501–3. The curtain wall was built by James III in 1467. At either end, east and west, it was flanked by a square tower. The east tower has been converted into the Three-gun Battery. The west tower still remains, with its oversailing battlement. The royal accounts show that this tower was built for James IV in 1496.

The principal buildings of the castle occupy the central or highest part of the rock. Behind them the Nether Bailey stretches to the north. The main buildings form a courtyard. On the south side is the Palace, a quadrangle round an inner close. It dates from the reigns of James IV and V, having been commenced in 1496, and is remarkable for the quaint mixture of Gothic and Renaissance details. Much of the embellishment is French, masons having been brought from that country. The windows retain their intersecting iron bars. Above them are the initials of James V. The interior has been greatly damaged, but retains some fine fireplaces. In the presence chamber was a magnificent oaken roof, with a series of carved heads of Scottish heroes.

Some are now preserved in the Smith Institute at Stirling and the National Museum of Antiquities, Edinburgh.

On the east side of the courtyard is the Hall, designed by the ill-fated favourite of James III, Thomas Cochran. It is noteworthy for its two enriched oriel windows. To the north is the Chapel Royal, built by James VI for the christening of Prince Henry in 1594. It has the Scottish "corbie steps," but the windows and doorway are classical. At the north-east corner of the upper rock is a pointed archway, grooved for a portcullis. This seems the oldest part of the castle, for the Exchequer Rolls of 1380–1 mention the building of a "barbican and north gate."

During the fourteenth century, when Scotland was impoverished by her struggle with the Plantagenets, the castle in vogue was a simple square tower, vaulted on one or more floors, and capped with an open battlemented parapet, within which rose a high roof between crow-stepped gables. This form of tower persisted even after improved conditions had led to the development of superior types of dwelling. An early example of the simple tower is Mains Castle near Drymen. It is of great strength, the walls being 8 feet hick, and the first three floors vaulted. The entrance, reached by a ladder, is on the second floor. Other specimens of the square tower may be seen at Bardowie, Castle Cary, Culcreuch, and Touch. Bardowie Castle has a fine open-timber roof of the sixteenth century. Castle Cary, romantically placed on the steep bank of the Red Burn, is a sixteenth century tower with additions made in 1679. The iron "yett" still exists. Many of the stones in this castle have been taken from the Roman fort near by.

The first improvement on the square tower was the adding of a wing, giving the structure the shape of the letter **L**. This plan also persisted until a late period. Examples exist at Auchenbowie, Old Sauchie, and Stenhouse.

Castle Cary

With the introduction of fire-arms, flanking towers came into vogue. Duchray Castle is an example of this development. It consists of an oblong house with a round tower at one angle. This castle dates from the later sixteenth century, and formed the headquarters from which in 1653

Glencairn launched his rising against Cromwell's rule in Scotland.

The picturesque ruin of Torwood Castle, built in 1565, is a good example of a Scottish castellated mansion when military conditions had largely been forgotten. It is built on a modification of the L plan. The basement is well provided with shotholes, but otherwise there are no defensive features, and flanking towers or turrets are absent, though the baronial style is maintained in the stepped gables. The courtyard buildings and well may still be traced. Another type of late building is on the T plan, of which an example occurs at Old Leckie. This house preserves its iron "yett." At Airth Castle the old open "round" and the later roofed turret exist side by side. Carnock Castle, built in 1548 and enlarged in 1634, retains some fine plaster ceilings of the latter date.

From the fifteenth century onwards the tower-house plan, in its various forms, was sometimes abandoned for castles built round a courtyard. The Palace at Stirling Castle is an instance of this. Examples may also be seen at Mugdock and Duntreath. Little remains of Mugdock Castle except two square towers, a gateway, and a greatly ruined chapel, all remarkable for their excellent masonry. This castle belonged to the Marquis of Montrose, and was destroyed by his Covenanting enemies in 1641. Duntreath Castle has a fine turreted gate-house of the early seventeenth century, with folding doors of iron strips riveted on a frame, instead of the usual open "yett." The gate-house bears the family arms, with the unique peculiarity of a camel as a single supporter, placed below the shield.

18. Architecture—(iii) Municipal and Domestic.

About the middle of last century John Hill Burton wrote the following description of Stirling: "The town, clinging as it were to the edge of the castle rock, down which it decreases in irregular terraces, forms an harmonious cluster, of which precipices and towers form a predominant feature, with which the venerable and picturesque buildings on the slope harmoniously combine.... The various stages of the approach do not disappoint the estimate which the eye may have formed from some distant eminence.... The town is full of old houses, with paved courts and arched entrances, from which pleasant gardens, wherein we may notice the antiquity of the fruit trees, stretch down on either side of the descent crowned by the castle, and exhibit in considerable vitality the economy of the old Scottish towns, where the houses were huddled together on an eminence, while all round them the gardens of the citizens stretched fanlike to the sun and the pure air. Ascending the main street, steep and rough, the edifices on either side, generally of a venerable character, become more picturesque and baronial in their character, forming a gradation from the burgher's high-gabled dwelling to the imperial palace and fortalice."

Since these words were penned the march of modernism has shorn Stirling of many of its ancient features; but the town retains its quaint, old-world character, and still possesses some fine specimens of old Scottish domestic architecture. Perhaps the most interesting of these houses is

Mar's Work, at the head of Broad Street. The building presents a highly ornate front pierced with many windows. Two semi-hexagonal towers flank the arched entrance, which gives access by a "pend" into what had been a court. Over the arch are the royal arms of Scotland, while the towers show the arms of the Regent Mar and his wife,

Mar's Work, Stirling

Annabella Murray of Tullibardine, with the date 1570. On the towers are also the following two inscriptions, with a third one at the back of the building:

THE MOIR I STAND ON OPPIN HITHT
MY FAVLTIS MOIR SUBIECT AR TO SITHT.

I PRAY AL LVIKARIS ON THIS LVGING
VITH GENTIL E TO GIF THAIR IVGING.

ESPY SPEIK FVRTH AND SPAIR NOTHT
CONSIDDIR VEIL I CAIR NOTHT.

It has been suggested that the stones of this building
were plundered from Cambuskenneth Abbey, of which
Mar had received a grant from Queen Mary; and this
supposition is so far confirmed by a consecration cross on
the back of the building. But much of the ornament which
has been thought ecclesiastical is quite common in domestic
work of the period.

On the east side of Castle Wynd stands Argyll's
Lodging, now a military hospital. It has justly been called
"the finest specimen of an old town residence remaining in
Scotland." The house occupies three sides of a quadrangle,
of which the fourth side, towards the street, is closed by a
high wall with a handsome gateway. From the opposite side
of the court a pillared porch gives access to the house. In
the angles of the courtyard turret stairs with high pointed
roofs lead to the upper floors, which have ornamented
windows, crow-stepped gables, and carved chimneys. Many
of the rooms preserve their enriched fireplaces and other
ancient features. As appears by dates on the walls, this
house was begun in 1632 by Sir William Alexander of
Menstrie, afterwards Earl of Stirling (p. 125) and completed
by the Marquis of Argyll in 1674. Many distinguished
visitors have passed its threshold, including Charles II and
James II. It formed the headquarters of the Duke of Argyll
during the "Fifteen," and of the Duke of Cumberland
during the "Forty-five."

Near Stirling Church is Cowane's Hospital, erected in
1637 from funds gifted by John Cowane, Dean of Guild.

With its dignified steeple and a statue of the founder, it is
a good example of the plainer Scottish domestic style. It
contains some valuable relics, including John Cowane's

Argyll's Lodging

chest of solid oak, dated 1636, and embellished with scrip-
tural texts which illustrate the benevolent character of the
owner. Attached to the house is a fine old garden, with a
sundial dated 1661.

The modern buildings of Stirling are of no importance, but mention may be made of the Tolbooth, built in 1701, with its picturesque spire. Portions of the old town wall still remain in the Back Walk. Stirling Bridge was erected between 1400 and 1415, previous structures having been of wood. It consists of four semicircular arches strengthened

Old Bridge, Stirling

by massive triangular abutments on the piers, and measures about 270 feet in length and 15 feet in width across the parapets. It is all built of hewn sandstone. The south arch was rebuilt in 1749, the original arch having been blown up during the "Forty-five" to prevent the Highlanders entering the town.

The ancient burgh of Falkirk has been completely modernised, but retains the steeple of its old Town Kirk, in the vestibule of which are preserved two pairs of effigies of a knight and his lady, said to be some of the early lords of Callendar. In the kirkyard is the reputed grave of Sir John de Graeme, killed at the battle of Falkirk (1298). Three successive stones have been superimposed on the original monument, which shows the wasted effigy of a knight.

The shire contains many mansions, some of great size, but few possess any claims to architectural importance. Kinnell House and Dunipace House are fine examples of the classical edifice, symmetrically planned. An example of the revival of the Scottish style in modern times may be seen at Kinnaird House. In Glenbervie House is a noted staircase of Spanish chestnut grown on the estate. One of the largest mansions in the shire is Callendar House, which contains some ancient work but is mainly a modern edifice in a mixed Scottish and classical style. Another very large building is Lennox Castle, an embattled edifice rendered imposing by its mere mass and height. Buchanan Castle is noted for its finely timbered park and for the magnificent views of mountain scenery which it commands. Other notable modern mansions are Polmaise Castle, Leckie House, and Airthrey Castle. Near Larbert are two extensive piles of building—the Stirling District Lunatic Asylum and the Scottish National Institution for the Training of Imbecile Children.

The Wallace Monument, erected in 1861–9 as a national memorial, is a striking pinnacled tower, 220 feet in height, crowning Abbey Craig. It contains a gallery of busts of famous Scotsmen and a collection of ancient arms and armour, and documents concerning the career of Wallace.

Good building material is obtained from the numerous sandstones of the Old Red and Carboniferous formations,

Airthrey Castle

and the intrusive basaltic and doleritic masses supply excellent paving stones and road-metal. Most of the medieval buildings in Stirling were erected in sandstone from the ancient quarries at Ballengeich and Raploch. Slates have to be imported. Prior to the war, most of the slates used in

Stirlingshire came from Aberfoyle, Easdale, and Balla-chulish. The yellow clay of the Carse of Stirling makes excellent bricks and tiles, but this is now almost an extinct industry. Tile drains for agriculture are still made.

Wallace Monument, Stirling

19. Communications, Past and Present.

From the earliest times the inhabitants of Scotland must have possessed their hunting paths and trade routes through the primeval marshes and forests. Doubtless the course of these has to a large extent been perpetuated by the later roads. The Romans were great road engineers, and their advent into Stirlingshire is marked by the introduction of military highways, some traces of which still remain. From York, the capital of the British Province, the trunk route ran by Aldborough and Catterick to the Wall of Hadrian near Corbridge. Passing through the Wall, it traversed the moorlands of Northumbria, crossed the Tweed at Newstead, struck the Forth at Inveresk, touched the east end of the Antonine Wall at Carriden, and thence westward to Falkirk formed part of the military way which ran from sea to sea in rear of the Wall (see p. 81). A mile or two beyond Falkirk the road struck out north by Camelon, crossed the Forth at or near Stirling, and held onward to the great forts at Ardoch and Strageath in Perthshire. Another road is supposed to have branched off at Carluke from the western trunk route by Carlisle, Birrens, and Annandale, and to have joined the Wall at Castlecary.

After the withdrawal of the Romans their splendid road-system fell into decay. Throughout the Middle Ages, and indeed until the eighteenth century, the communications of Scotland were in a wretched condition. Most of the roads were mere tracks—"green ways," they are significantly termed—but others, referred to as "king's ways," "highways," and "causeways," must obviously have been of a better type. The "king's way" was in the royal peace, and crimes committed on it were severely punished. There was little wheeled traffic, and trade was mainly carried on by means of pack-horses. Bridges were few and far between, but as early as the thirteenth century the Forth was crossed by a wooden bridge at Stirling. Doubtless there were ferries at a still earlier date.

A famous early bridge builder was Robert Spittal, who died in 1558. He was tailor to King James IV and his Queen, and besides endowing a hospital which still exists in Stirling, his public spirit took the form of bridge-building. Bridges over the Bannock at Bannockburn, the Devon at Tullibody, and the Teith near Doune, all still standing, were erected at his expense.

The present system of roads in the county has been evolved since the passing of the Turnpike Roads Act of 1751. At present Stirlingshire is well served by excellent roads. The main north road from Edinburgh runs by Falkirk and Larbert to Stirling. From Falkirk a good road runs westward to Dennyloanhead, whence Glasgow may be reached either directly by Castlecary and Cumbernauld, or by Kilsyth and Kirkintilloch. A branch from the latter road ascends the Glazert Water by Milton to Strathblane. From Glasgow a road runs northward by Milngavie to Strathblane, or by a left branch—known as the Stockiemuir road—to Drymen, from which there is a passable road up the shore of Loch Lomond to Rowardennan, and a good road to Balloch and the Leven valley. The north road from Glasgow proceeds from Strathblane across the shire to Aberfoyle, sending off an eastward branch by Killearn, Balfron, and Buchlyvie, and so down the Forth valley through Kippen and Gargunnock to Stirling. The central uplands of the county are traversed by a second-class road from Campsie to Fintry, known as the Craw Road, with branches from Fintry eastward down the Carron to Denny, and westward down the Endrick to Killearn. From Stirling a road leads eastward down the Forth to Airth, and so on to join the Edinburgh to Falkirk road at Polmont. Slamannan in the south-east upland district is connected by good roads with Falkirk, Airdrie, and Bathgate. The extreme north-western corner of the shire, beyond Ben Lomond, is traversed by a first-class road from Inversnaid on Loch Lomond through Glen Arklet to Stronachlachar on Loch Katrine. There are numerous district roads of good quality in every part of the shire.

Stirlingshire is traversed by the Forth and Clyde Canal from its eastern terminus at Grangemouth to Castlecary. The project of connecting the Forth and Clyde by a cutting across the narrowest part of Scotland was mooted as far back as the reign of Charles II, but work was not begun until 1768, and the canal was opened for traffic only in 1790. The cost of the undertaking was partly defrayed by Government out of the revenues of confiscated Jacobite estates. The total length of the canal is $38\frac{3}{4}$ miles, its depth about 10 feet, and its breadth 56 feet. There are 39 locks, and the greatest height, 156 feet, is reached at Castlecary. The canal admits vessels of 68 feet keel, 19 feet beam, and $8\frac{1}{2}$ feet draught. At Portdownie, near Falkirk, it is joined by the Union Canal from Edinburgh, which enters Stirlingshire near Muiravonside. Constructed in 1818–22, the Union Canal is $31\frac{1}{2}$ miles long, 40 feet wide, and 5 feet deep. It was never a commercial success, and both canals were ruined by the railway. In 1848 the Union Canal was purchased by the Edinburgh and Glasgow Railway, afterwards amalgamated with the North British (now London and

North-Eastern) Railway Company. The Forth and Clyde
Canal was bought in 1867 by the Caledonian (now London,
Midland and Scottish) Railway Company. Both canals are
still in use for local purposes, but are of little wider importance.

The disposition of the Stirlingshire railway system is
dictated by the great mass of elevated, barren country,
almost empty of population, which fills up the centre of
the county. Both the great railway companies in Scotland,
the London and North-Eastern and the London, Midland
and Scottish, have lines in Stirlingshire, the former representing the old North British, and the latter the old
Caledonian Companies respectively. The North British
Railway was formed in 1865 by the amalgamation of the
old Edinburgh to Berwick Railway with the Edinburgh
and Glasgow Railway, which was opened in 1841, and
had previously absorbed the Monkland and Kirkintilloch,
the Ballochney, and the Slamannan Railways, of which
the last mentioned, opened in 1840, was the first railway
line in Stirlingshire. The Forth and Clyde line from Balloch to Stirling was constructed in 1854, and leased to the
North British Company in 1871. The Caledonian Railway,
completed in 1848, comprised a main line from Carlisle
to Carstairs, branching thence westward to Glasgow and
eastward to Edinburgh. The old Scottish Central Railway,
between Greenhill and Perth, opened in 1848, added in
1859 a branch westward to Denny, and the whole was
taken over by the Caledonian Company, after prolonged
litigation, in 1865. The Caledonian and North British
Railways were merged respectively in the London, Midland

and Scottish and London and North-Eastern lines as a
result of the Railways Act of 1922.

The trunk line of the London and North-Eastern Railway from Edinburgh to Glasgow enters Stirlingshire west
of Linlithgow, and runs by Falkirk (High Station) along
the Bonny Water, passing out of the shire just beyond
Castlecary station. At Polmont Junction a branch line
holds to the right, and passing through Falkirk (Grahamston Station) joins the trunk line of the London, Midland
and Scottish Railway at Larbert Junction. Along this line
the London and North-Eastern Company has running
rights as far as Stirling. From Stirling the old Forth and
Clyde line ascends the Forth valley by Gargunnock and
Kippen to Buchlyvie, from which a branch is sent out
north-westward to Aberfoyle beyond the shire boundary.
From Gartness on the Endrick the main line, passing out
of the shire where it crosses the river, holds westward to
Balloch at the foot of Loch Lomond. At Gartness a branch
turns southward and, crossing the Endrick, ascends the
Blane valley, traverses the watershed between Strathblane
and Campsie, descends the valley of the Glazert, and,
leaving the shire near Kirkintilloch, joins the Edinburgh
and Glasgow line at Lenzie in Dumbartonshire.

From Larbert Junction a line runs along the north bank
of the Bonny Water by Dennyloanhead to Kilsyth, and
thence descends the valley of the Kelvin to its junction
with the Allander Water south-east of Bardowie Loch,
where it passes out of the county on its way to Glasgow.
From Stirling a line runs eastward from Causewayhead to
Alloa beyond the shire boundary. The south-eastern corner

of the shire is served by the Slamannan and Bo'ness section of the railway, which at Blackstone Junction gives off a branch to Bathgate in Linlithgowshire.

The trunk line of the London, Midland and Scottish Railway from Glasgow to the north enters Stirlingshire at Castlecary and runs by Greenhill Junction to Larbert Junction, whence it holds northward by Plean and Bannockburn to Stirling, passing out of the shire across the Allan Water on its way to Dunblane, Crieff, and Perth. The London, Midland and Scottish Railway has also running rights on the London and North-Eastern line from Larbert Junction eastward by Falkirk and Linlithgow to Edinburgh. Branch lines of this railway also give access from Falkirk (Grahamston Station) to Grangemouth, from Larbert Junction to Denny, and from Plean across the Forth to Alloa. At Dunmore Moss a short branch of the last line, used for goods traffic only, connects South Alloa on the Stirlingshire side of the river.

Both railways maintain a joint service of steamers on Loch Lomond. Starting from Balloch Pier, the vessels run thrice daily during the season, calling at Balmaha, Rowardennan, and Inversnaid on the Stirlingshire side.

The Forth is a navigable river as far west as Stirling, where there is a small harbour. The traffic on the river, at one time considerable, is now of little account. Before the war there was a service of passenger steamers from Leith to Stirling, but it has not been resumed above Alloa.

20. Administration and Divisions.

It has already been explained (Chapter 1) that the division of Scotland into shires or counties and parishes was the work of the Anglo-Norman immigrants during the twelfth and thirteenth centuries, and that Stirlingshire took origin in the administrative area dependent on its royal castle. The power of the feudal nobility led in the fourteenth century to the office of sheriff becoming hereditary; but in 1747, after the last Jacobite rebellion, appointments to this office were vested in the crown. The sheriff had jurisdiction over "pleas of the crown," but there existed also wide feudal jurisdictions in the courts of barony and regality controlled by the barons and prelates. Over the sheriffs in early times were the four Justiciars of Galloway, Lothian, and the districts north and south of the Mounth. The sheriff also accounted to the Exchequer for the revenues of his county, and led its levies to the national army. The town of Stirling, erected into a royal burgh early in the thirteenth century, obtained thus a substantial control over its own revenues and government. There was also the King's Lordship of Stirling, attached to the royal castle. It was given as part of her dowry to Queen Margaret of Denmark, by James III, and later came into the possession of the Earls of Mar, who were hereditary governors of the castle.

Nowadays the administrative unit is the County Council, charged with the general administration of the shire, including the raising of county revenues by rating and borrowing, the enforcing of the Diseases of Animals and

Weights and Measures Acts, and various other functions. The County Council has the sole responsibility for the finance of County Administration. Over it presides the County Convener, assisted by a Vice-Convener. County Councillors hold office for three years, and all retire at the same time. The official head of the County is the Lord

County Council Buildings, Falkirk

Lieutenant, under whom are a Vice-Lieutenant and a number of Deputy Lieutenants and Justices of the Peace.

Under the County Council are the District Committees, which are the local authority for administering the Public Health Acts and maintaining roads and bridges, but have no power to raise money. Stirlingshire has three District Committees, Eastern, Central, and Western. The District

8-2

Committees consist of the County Councillors for the District, Parish Councillors nominated by each Parish Council in the District, and representatives from burghs where road management has been transferred to the District Committee. Each parish has thus two representatives on the District Committee, one elected by the voters and the other appointed by the Parish Council, with the burgh representatives in addition. The Parish Council looks after the Poor Law, and (except in burghal parishes) administers the Vaccination Acts, appoints registrars, cares for churches and manses, provides burial grounds, and maintains rights of way. The number of Parish Councillors in landward parishes is fixed by the County Council, and in burghal parishes by the Town Council.

Each burgh has a Town Council consisting of Provost, Magistrates, and Councillors, holding office for three years. One-third of these retire annually. The Town Council is charged generally with the administration of the burgh : it is the local authority for public health, looks after the streets, public buildings, and cleansing, and controls various municipal undertakings such as water-works, tramways, lighting, slaughter-houses, parks and recreation grounds, and harbours. The Magistrates are elected annually by the Council. With the police-judges—who are ex-magistrates specially nominated by the Town Council—they form the police court for minor offences, and are also the licensing authority. There are six burghs in Stirlingshire, namely Stirling (a royal and parliamentary burgh), Falkirk (a parliamentary burgh), and the four police burghs of Grangemouth (now a parliamentary burgh), Kilsyth, Denny and Bridge of Allan.

The Scottish parishes seem to have originated in close dependence upon the manors or estates of the Norman intruders in the twelfth and thirteenth centuries. Whenever a Norman baron settled down upon his manor, there he erected a castle for himself and a church for his dependents. The church thus became the spiritual, as the castle was the civil, centre of the manor; and the manor became a parish under a priest responsible to the bishop of the diocese. Very often the lord granted the patronage of the parish church to some distant cathedral or monastery. Airth Church was thus "impropriated" to the Abbey of Holyrood. The association of the church with the castle is well seen at Airth, where church and castle exist side by side on the same hill.

Stirlingshire has 22 civil parishes, 7 partly burghal, 15 entirely landward: Airth, Baldernock, Balfron, Buchanan, Campsie, Denny, Drymen, Dunipace, Falkirk, Fintry, Gargunnock, Grangemouth, Killearn, Kilsyth, Kippen, Larbert, Logie, Muiravonside, St Ninian's, Slamannan, Stirling, and Strathblane. There are 28 Registration Districts, 7 partly burghal and 21 entirely landward.

For judicial purposes Scotland is divided into 4 Circuits and 15 Sheriffdoms. The Sheriffdom of Stirling, Dumbarton, and Clackmannan is in the Western Circuit. For sheriff court purposes the county of Stirling is divided into two districts—the western with its court at Stirling, and the eastern with its court at Falkirk. There are 61 Police Districts in Scotland: Stirling County and Burgh each form such a District, each maintaining its own constabulary.

Scotland is divided into 26 Lunacy Districts, for each of which there is a Lunacy Board. The Stirling Lunacy

District comprises Clackmannan, Dumbarton, Lanark (including part of New Kilpatrick within the boundary of Glasgow), Stirling and West Lothian (including Kirkliston Parish, which is partly in Mid Lothian). Stirling, Falkirk, and the shire each form a separate National Health Insurance area. Falkirk, Grangemouth, Kilsyth, and Stirling are each Burgh Licensing Districts, and the shire forms a County Licensing District. Falkirk and Stirling have each a separate Licensing Court and a separate Appeal Court: Grangemouth and Kilsyth have each a separate Licensing Court, and together have a joint Appeal Court. Denny and Bridge of Allan are merged in the county for licensing purposes. The shire has 21 special Water Districts, 15 special Lighting Districts, 19 special Drainage Districts, and 12 special Scavenging Districts. The Stirlingshire and Falkirk Water Board administers the water-supply of the Eastern and Central Districts of the county, and the Burgh of Falkirk. The Stirling Waterworks Commissioners are the water authority for the Burgh of Stirling, the landward part of the parish of Stirling and parts of the parishes of St Ninian's and Logie. The reservoir at North Third, on the Bannock Burn, is owned under a partnership agreement by the Stirling Waterworks Commissioners and the Town Council of Grangemouth.

Under the Education Act of 1872 each parish elected a School Board, but the Act of 1918 has now established an Education Authority for the whole county, with its offices at Stirling, to control alike primary, intermediate, secondary, and continuation schools. Under the Act, education is compulsory between the ages of 5 and 15 years, but in practice it has not yet been found possible to raise the age beyond the previous limit of 14 years.

Under the Representation of the People Act, 1918, by which the Scottish constituencies were re-grouped, Stirlingshire is included in three parliamentary constituencies, each returning a member: Stirling and Falkirk (comprising the burghs of Stirling, Falkirk, and Grangemouth); Clackmannan and Eastern Division; and Western Division.

Ecclesiastically Scotland is divided into 16 Synods, below which are Presbyteries consisting of groups of parishes. The Synod of Perth and Stirling includes the Presbyteries of Dunkeld, Weem, Perth, Auchterarder, Stirling, and Dunblane. The ecclesiastical parishes do not always coincide with the civil parishes. Thus the ecclesiastical parish of Cumbernauld is partly in Dumbartonshire and partly in Stirlingshire, whereas the civil parish belongs wholly to the former. The ecclesiastical parish of Logie includes part of Clackmannanshire, the civil parish being wholly in the County of Stirling. Stirlingshire contains the *quoad sacra* parishes of Bannockburn, Banton, Bonnybridge, Bothkennar, Bridge of Allan, Buchlyvie, Camelon, Grahamston, Haggs, Kerse, Larbert and Dunipace, Laurieston, Marykirk, Plean, Polmont, Shieldhill and Blackbraes, and Stenhouse. The *quoad sacra* parish of Gartmore is partly in Perthshire and partly in Stirlingshire. Moreover, the ecclesiastical parishes are by no means all included in the Presbytery of Stirling, for the Presbyteries of Dunblane, Linlithgow, Dumbarton and Glasgow all include Stirlingshire parishes.

21. The Roll of Honour.

In the roll of honour of Stirlingshire there are at least two names of quite outstanding eminence—George Buchanan, the humanist, and James Bruce, the Abyssinian explorer.

George Buchanan

George Buchanan, descended from the Buchanans of Drumnakill, was born at the farm of Mid Leowen in the parish of Killearn in 1506. He was educated at the Universities of St Andrews and Paris, and in 1529 was appointed

Professor of Latin in the College of Sainte-Barbe. On his return to Scotland in 1537 he took the field in a literary campaign against the corruption of the Roman Church. James V made him tutor to one of his natural sons, but the resentment of the priesthood soon forced him to retire to the Continent. After various wanderings in France and Italy he returned to Scotland in 1561. Despite his Protestantism he was chosen to be her classical tutor by Queen Mary, who read Livy with him daily after dinner. In 1566 he became Principal of St Leonard's College, St Andrews. Notwithstanding the favour that he had received from Queen Mary, Buchanan after Darnley's murder joined the party of the Earl of Moray, and smirched himself by coarsely calumnious attacks upon his benefactress. In 1570 he became tutor to the young King James VI, and the severity with which he disciplined his royal pupil in Stirling Castle has become proverbial. In 1571 he was appointed Lord Keeper of the Privy Seal. He died at Edinburgh in 1582. Despite grave faults as a man, Buchanan's vast learning has earned for him a world-wide reputation as one of the most brilliant scholars of the Renaissance. His writings in poetry and prose are voluminous.

James Bruce was born at Kinnaird House in 1730. Travels in Portugal and Spain fired his interest in Arabic literature and Oriental studies generally. Forming the ambitious resolve to discover the sources of the Nile, in June 1768 he went to Alexandria, obtained the support of the Mameluke, and after various adventures, including a visit to Arabia, in February 1770 reached Gondar, the capital of Abyssinia. By his frankness, courage, and variety of accomplishments he won the favour of the Abyssinian ruler and people, among whom he stayed for two years. In November 1770 he reached the source of the Blue Nile. In 1790 he published in five volumes his *Travels to Discover the Source of the Nile*. At first his work was received with some disbelief, but subsequent exploration has confirmed its accuracy. He died at Kinnaird House in 1794.

At least one King of Scotland, James III, was born in Stirling Castle (1451), which also was the birthplace, in 1594, of Prince Henry, eldest son of James VI.

To the ranks of British statesmen and soldiers Stirlingshire has contributed no names of outstanding eminence, but not a few of distinction. Foremost among the list of statesmen comes the name of Sir Henry Campbell-Bannerman (1836–1908), Liberal M.P. for the Stirlingshire Burghs from 1868 to his death, and Prime Minister from 1905 till within a few weeks of that event. Although born in Glasgow, Sir Henry was of a Stirlingshire stock, his father having migrated from the western portion of the county in early life. Sir Henry's ministry will always be remembered for its bold gift, so shortly after the Boer War, of a self-governing constitution to the Transvaal and Orange River Colonies. Among naval and military leaders, we may mention the names of Lieutenant-General William Baillie of Letham, who commanded the Scottish infantry with great distinction at Marston Moor in 1644, but next year was defeated by Montrose at Alford and Kilsyth; Major-General Thomas Dundas of Carron Hall (1750–94), who led a brigade in the American War of Independence, was one of the Commissioners who negotiated the surrender at York Town,

Virginia (1781), and took a prominent part in the capture of Martinique, St Lucia, and Guadeloupe from the French

James III of Scotland

in 1794: General Sir David Russell (1809–84), who took part in the capture of Lucknow during the Indian Mutiny: Sir John Downie, who served under Wellington in the

Peninsular War, became a brigadier in the Spanish Army and was governor of Seville; Admiral Sir Charles Napier (1786–1860) who commanded the British fleet in northern waters during the Crimean War; Captain Robert Spottiswood of Dunipace, who with his ship the *Lord Nelson* (26 guns) fought a gallant but unsuccessful duel against a French privateer, the *Bellona* (34 guns) in 1803; and Admiral Viscount Keith (1746–1823), who took a prominent part in Lord Hood's defence of Toulon in 1793, directed the naval wing of the expedition which, by capturing Capetown in 1795, laid the foundations of British dominion in South Africa, and was in command of the fleet that landed Sir Ralph Abercrombie in Egypt in 1801. Sir Ralph himself was a Stirlingshire landowner, his family being the proprietors of Airthrey estate near Stirling.

Among lawyers connected with Stirlingshire may be noted Sir Robert Spottiswood of Dunipace (1596–1646), who was President of the Court of Session under Charles I and was executed by the Covenanters; his successor at Dunipace, Sir Archibald Primrose (1616–79), Lord Justice-General under Charles II, an ancestor of the Earl of Rosebery; and Professor John Erskine (1695–1768), author of the *Institutes of the Law of Scotland*.

In the realm of literature, apart from Buchanan Stirlingshire has produced few men of note. But it can boast the name of Robert Henry (1718–90), whose *History of England* enjoyed great vogue in its day. A more distinguished name is that of Professor Henry Drummond (1857–97), theologian, philosopher, scientist, traveller, and philanthropist, whose *Natural Law in the Spiritual World* and *Ascent of*

Man created an enormous sensation when published, but who is best remembered now as a teacher and inspirer of young men. Closely connected with Stirlingshire, though not a native, was Sir William Alexander of Menstrie (*circa* 1567–1640), Viscount Canada and Earl of Stirling, statesman, poet, and friend of Drummond of Hawthornden—best known, perhaps, to-day for his connexion with James VI's ill-starred attempt to found a colony in Nova Scotia. He built the oldest part of Argyll's Lodging, Stirling, in 1632–5, and is buried in Stirling Church. Mention may also be made of George R. Gleig, the friend of Keats, and author of *The Subaltern* and other works, who was a native of Stirling, and of Dr John Moore, who was born at St Ninian's, the writer of *Zeluco* and many other novels, but better known as the father of Sir John Moore, the hero of Corunna. In art may be noted the names of Sir George Harvey and Sir Daniel Macnee, successive Presidents of the Royal Scottish Academy, and both natives of the shire.

Among men of science native to the county the only name of note is that of James Stirling (1692–1770), mathematician, who was born at Garden in Kippen parish, became Professor of Mathematics at Venice, and conducted about 1750 the original survey of the Clyde estuary, as a result of which Glasgow has grown into a seaport. But it should be remembered that John Napier (1550–1617), the inventor of logarithms, was closely associated with Stirlingshire. He owned property in the parish of Drymen, and here at Gartness House much of his epoch-making *Mirifici Logarithmorum Canonis Descriptio* was written.

It may be mentioned that Henry Guthrie, Bishop of

John Napier

Dunkeld and author of *Memoirs of Scottish Affairs during the Civil War*, and James Guthrie, the famous Covenanting leader, executed in 1661, were successively ministers of Stirling. Ebenezer Erskine (1680–1754), founder of the Secession Church, was minister of the West Kirk. Robert Montgomery, whose promotion to the Bishopric of Glasgow gave rise to one of the most notorious cases in the history of the Church of Scotland, was minister of Stirling from 1572 to 1581.

22. The Chief Towns and Villages in Stirlingshire.

(The figures in brackets give the population in 1921, an asterisk denoting parishes. The figures at the end of each paragraph refer to the pages in the text.)

Airth (* 1777) is a village on the south bank of the Forth, 8½ miles south-east of Stirling. Anciently the place was one of considerable importance, and in 1511 James IV constructed here a harbour and dockyard for his new fleet. The outlines of the docks may still be traced on the level reclaimed land stretching northward to the Forth, which now flows between embankments at a distance of half a mile from the village. Picturesquely embosomed amid orchards and gardens and stately trees, Airth is now a delightful and deeply interesting example of a medieval village still retaining much of its ancient character, with its narrow, irregular lanes, market cross, and quaint old houses having corbie-stepped gables and heraldic enrichment. Immediately adjoining are Airth Castle and in its grounds the ruined Transitional parish church. (pp. 43, 59, 89, 99, 117.)

Baldernock (* 763) is a small village at the extreme south-west corner of the shire, 7 miles north of Glasgow. Near it are the Auld Wives' Lifts and Bardowie Castle. (pp. 30, 97.)

Balfron (* 1189), a village on the right bank of the Endrick, was founded as a centre of the cotton weaving industry in 1789, but has never recovered from the loss of business caused by the introduction of machinery, which concentrated the industry in larger and more accessible places. The village is finely situated, facing the Campsie Fells across Strathendrick, while behind are the Grampians, with Ben Lomond and Ben Venue. (pp. 20, 25, 59.)

Bannockburn is known to all the world as the scene of the decisive battle in the War of Scottish Independence : but the town, situated 2½ miles south-south-east of Stirling, on either side of the burn to which it owes its name, is a place of some consequence in itself, having collieries, carpet factories, and tanneries. (pp. 5, 10, 17, 59, 62, 65, 70, 72, 109.)

Bridge of Allan (3579) is a famous health resort on the north side of the Forth opposite and about 3 miles distant from Stirling.

The town is finely situated amid beautiful scenery on the left bank of the Allan Water, and is an excellent centre for climbing the southern summits of the Ochils. The town is a police burgh, and contains an art and science museum, reading room, Turkish baths, and excellent hotel and private accommodation of all kinds. Paper-making, weaving, and dyeing are the only important industries. Near the town are the Airthrey mineral wells, famed for their

Allan Water, Bridge of Allan

medicinal qualities. Airthrey Castle, dating from 1791 and surrounded by stately grounds, is in the vicinity. On the top of Abbey Craig is the Wallace Monument. (pp. 4, 6, 27, 28, 58, 59, 72, 75–6, 105–6, 124.)

Buchlyvie is a small burgh of barony on the west side of Kippen parish, 4 miles north-north-east of Balfron. To the north-west is all that remains of the once extensive Flanders Moss. (pp. 6, 15, 109, 112, 119.)

Cambusbarron, a village 1½ miles south-west of Stirling, is occupied with the woollen industry. (p. 73.)

Cambuskenneth, a haunt of artists, is a pretty little village picturesquely situated within a loop of the River Forth, connected by a ferry with Stirling. Adjoining is the ruined abbey containing the tombs of James III and his queen. (pp. 64, 65, 66, 89–91, 102.)

Denny-and-Dunipace (5130) is a police burgh formed in 1876 by the union of its two constituent parts, situated respectively on the right and left banks of the Carron, and connected by a handsome bridge. The town is 5½ miles west-south-west of Falkirk. Coal mining and iron founding are the two chief industries, but there are also chemical, paper, and engine works. In the neighbourhood are the famous mounds of Dunipace, Herbertshire Castle (an ancient pile recently destroyed by fire), and Dunipace House. (pp. 6, 10, 26, 28, 30, 40, 49, 58, 59, 75, 105, 111, 124.)

Drymen (*1214) is a quaint, old-world village, picturesquely situated near the north banks of the Endrick Water about 3 miles above its mouth in Loch Lomond. The village is within easy reach by road or rail from Glasgow, and is a popular centre for visiting Loch Lomond and the Highland district of Stirlingshire. Near it is Duchray Castle; and in the house of Gartness, on the Endrick, John Napier partly composed his great work on logarithms. (pp. 4, 17, 34, 87, 97, 98, 125.)

Falkirk (33,312) (Fawkirk, *Varia Capella, Eaglais Bhreac*), the largest town in Stirlingshire, is a municipal and police burgh situated about 11 miles south-east of Stirling, and half-way between Edinburgh and Glasgow. The town is well situated on rising ground overlooking the Carse of Falkirk. It includes the districts of Grahamston and Bainsford, Camelon, and Laurieston (anciently Langtoune). The principal buildings are the town church, town hall, burgh and county buildings, Dollar Free Library, and Camelon Hospital. On the outskirts of the town are the estate and mansion of Arnotdale, gifted to the community by Mr Robert Dollar, of San Francisco, a native of Falkirk. Here a golf course and tennis courts have been laid out, and a museum formed in the house. There are three fine public parks, the Victoria Park, the Princess Park, and the Dawson Park. From Falkirk High Station, at the south end of the town, a magnificent view may be had over the surrounding country, embracing the carse-land, the Forth valley, the Ochils, and the Wallace monument on Abbey Craig. Falkirk lies in the heart of the Stirlingshire coal-field, and is the centre of the light casting trade in Scotland. There are now nearly 30 foundries in the town. Other industries are the manufacture

Falkirk

Callendar House, Falkirk

of chemicals and explosives, brewing, distilling, flour-milling, and tanning. At Stenhousemuir, to the north-west, cattle sales are held in August, September, and October, but have latterly declined in importance. The Forth and Clyde Canal passes to the north, and the Union Canal to the south, being here carried through a tunnel 700 yards long. Historically Falkirk is a place of high interest. The Roman station of Camelon lies to the west, the Antonine Wall passes through the town, and in the neighbourhood were fought the two battles of Falkirk (1298 and 1746). To the south-east is Callendar House, captured by Cromwell in 1650. (pp. 6, 10, 27, 49, 50, 58, 59, 62, 65, 70, 71, 72, 74, 80, 82–4, 104–5, 108, 110, 112, 113, 115, 118, 119.)

Fintry (*381) is a small village in the central part of the shire, situated in upper Strathendrick, 17 miles north-east of Glasgow. The scenery in the valley is soft and pleasing, while the enclosing hills are barren, often rocky, and picturesque in outline. In the Dun Hill, near the village, is a striking basaltic colonnade. About 3½ miles to the east are the ruins of Sir John de Graeme's Castle. (pp. 11, 13, 93, 109.)

Gargunnock (*586) is a small village about a mile south of the Forth and 3½ miles east of Kippen. The Peel of Gargunnock is an early Norman earthwork. (pp. 10, 11, 13, 26, 37, 59, 109, 112.)

Grangemouth (9699), the seaport of Stirlingshire, is a police burgh situated on the Firth of Forth at the mouth of the Carron. Its history dates from the opening of the Forth and Clyde Canal in 1789. Through its large docks a great volume of commerce passes, the principal exports being coal, iron, and agricultural produce, while the principal imports are timber, pig-iron, and iron-ore. The industries include ship-building, iron-founding, and rope and sail making. Steamers ply between Grangemouth and London, Amsterdam, Rotterdam, Hamburg, and Christiania. There is a fine town hall, and a public park presented by the Marquis of Zetland, whose seat, Kerse House, adjoins the town. (pp. 5, 6, 13, 27, 43, 49, 50, 58, 59, 110, 113, 118.)

Killearn (*1052) is a village situated between the Blane and the Endrick about 1½ miles from their junction. At the farmhouse of Moss, in the neighbourhood, George Buchanan was born, and an obelisk 103 feet high has been erected to his memory in the village. (pp. 4, 14, 17, 109, 120.)

New Dock, Grangemouth

Kilsyth (7600) is a police burgh situated on the north bank of the Kelvin, 15 miles south-south-west of Stirling. The town is principally occupied with the coal mining and iron industries, but whinstone and sandstone are also quarried, and there are paper and cotton factories. Immediately north of the town rise the rugged treeless heights of the Kilsyth Hills, through which the Garrel Burn descends in a romantically picturesque, wooded gorge, in the vertical rocky sides of which the horizontal strata of coal may be seen. In the valley of the Garrel Burn above the town are the scanty remains of Kilsyth Castle, a former seat of the Livingstones of Callendar, destroyed by Cromwell in 1650. To the north-east of the town is the site of Montrose's victory over the Covenanters in 1645, but the battlefield is now largely inundated by a reservoir for the Forth and Clyde Canal. The present town was founded in 1665, and became a burgh of barony in 1826. (pp. 11, 13, 26, 27, 28, 31, 49, 50, 58, 68, 93, 109.)

Kippen (*1508) is a village on the Forth about 9 miles west of Stirling. In the beginning of last century it was an important whisky centre, but is now chiefly busied with its agricultural market. A mile west of the village is the site of the Kippen Conventicle, held in 1676, and at Ford of Frew, near the railway station, Prince Charles Edward lost his cannon during the retreat of 1746. (pp. 3, 25, 59, 71, 109, 112.)

Larbert (*12,389) is an important railway junction 8 miles south-east of Stirling. In the parish graveyard is a monument to James Bruce, the Abyssinian explorer. Coal mining is the principal industry. Near by are the Scottish National Institution for Imbecile Children and the Stirling District Asylum. To the north-west are the remains of Torwood Forest, with its broch and ruined castle. (pp. 10, 14, 27, 40, 50, 78, 99, 105, 109, 112, 113.)

Lennoxtown, situated about 11 miles north-east of Glasgow on the Glazert Water, is the centre of the Campsie mining and manufacturing district with its busy coal mines and limestone quarries, its chemical works, bleachfields, and printworks. The town consists of a single long street and is devoid of any distinctive buildings. (p. 58.)

Polmont is a village and railway junction near the eastern border of the shire. In the churchyard the historian Robert Henry is buried. (pp. 27, 71, 112.)

Slamannan (*3409) is a mining village in the south-eastern portion of the shire, on the right bank of the River Avon. (pp. 6, 12, 27, 32, 50, 61, 62.)

South Alloa is a small village and port in Airth parish, immediately opposite Alloa in Clackmannanshire. It has large timber yards and a steam ferry across the Forth. (p. 112.)

Stirling (21,345), the county town, is situated on the right bank of the Forth, 39¾ miles north-west from Edinburgh, and 29½ miles north-east from Glasgow. The ancient burgh occupies the slopes of

Smith Institute, Stirling

the "tail" extending southward from the castle rock : the new town has expanded fan-like over the southern plain. Despite much modernisation, the old town retains the character and atmosphere of a medieval Scottish burgh. Its most interesting buildings are the Castle, the Church, Mar's Work, Argyll's Lodging, Cowane's Hospital, and the Old Town House. The burgh was surrounded by a wall with two entrances, one by the old bridge, and the other by the South Port, a little west of the present line of Port Street. Among modern places and buildings of interest in the town are the Cemetery next the Castle, with the Covenanters' Memorials : the old Burgh Buildings, with a statue of Sir William Wallace ; the modern

Municipal Buildings ; the restored Town Cross ; the Smith Institute, founded in 1873 by T. S. Smith, a local artist, and containing an excellent Museum and Art Gallery ; the Royal Infirmary ; and the High School. Below the ancient bridge are the modern road bridge, erected in 1829 by Robert Stevenson, and the railway viaduct. There is a small harbour, accessible only at high tide. The chief industries

Thomas Stuart Smith

are the manufacture of carpets and hosiery, iron-founding, carriage building, brush making, and the manufacture of rubber goods, furniture, and agricultural implements. Stirling has been a royal burgh from the reign of Alexander II ; and its annals, with those of its famous castle, are closely entwined with the national history of Scotland. To the west of the town is the beautiful King's Park ; the surroundings are full of interest and charm, and the fine outlines of the Ochils and the Grampians lend character to the scenery. From the ramparts of the castle one of the grandest views in Scotland may be enjoyed. Northward, across the Forth, are Cambuskenneth Abbey and the Wallace Monument; while 2 miles to the south is the battlefield of Bannockburn. (pp. 2, 3, 4, 5, 15, 28, 29, 36, 40, 49, 50, 58, 59, 63–70, 71, 91–2, 93, 94–7, 100–4, 106, 109, 110, 111, 112, 113, 114, 118, 122.)

St Ninian's (* 14,832), situated on the main south road about a mile outside Stirling, is really a continuation of that town, but retains much of its ancient independent character in its quaint old houses with their crow-stepped gables and the lofty tower of its ruined church, destroyed by the Jacobites in 1746. It has woollen industries, candle works, and nail factories. Immediately to the south and east is the battlefield of Bannockburn. (pp. 50, 58, 62, 63, 87, 92–3.)

Strathblane (* 1275) is a village situated on the River Blane, in the south-west portion of the shire, about 11 miles north-west of Glasgow. At Blanefield, close by, are important printworks. (pp. 14, 20, 36, 109.)

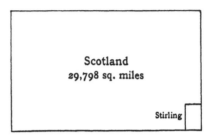

Fig. 1. Area of Stirlingshire (451 sq. miles) compared with that of Scotland

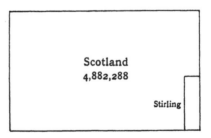

Fig. 2. Population of Stirlingshire (161,726) compared with that of Scotland in 1921

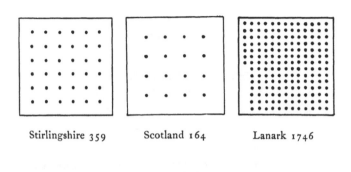

Stirlingshire 359   Scotland 164   Lanark 1746

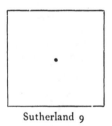

Sutherland 9

Fig. 3. Comparative Density of Population to the square mile in 1921

(*Each dot represents 10 persons*)

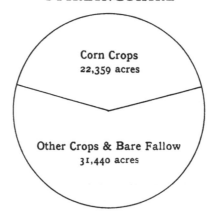

Fig. 4. Proportionate area under Corn Crops compared with that of other cultivated land in Stirlingshire in 1924

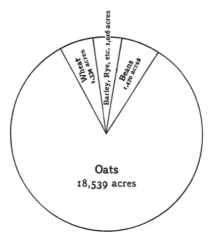

Fig. 5. Proportionate areas of chief Cereals in Stirlingshire in 1924

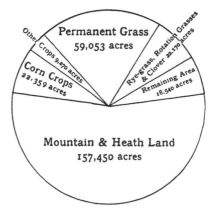

Fig. 6. Proportionate areas of land in
Stirlingshire in 1924

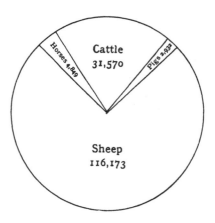

Fig. 7. Proportionate numbers of Live Stock in
Stirlingshire in 1924